THE ILLUSTRATED HISTORY OF THE
CHERRY MINE DISASTER
of 1909

JIM RIDINGS

America Through Time is an imprint of Fonthill Media LLC
www.through-time.com
office@through-time.com

Published by Arcadia Publishing by arrangement with Fonthill Media LLC
For all general information, please contact Arcadia Publishing:
Telephone:843-853-2070
Fax:843-853-0044
E-mail: sales@arcadiapublishing.com
For customer service and orders:
Toll-Free 1-888-313-2665

www.arcadiapublishing.com

First published 2020

Copyright © Jim Ridings 2020

ISBN 978-1-63499-202-2

All rights reserved. No part of this publication may be reproduced, stored in a retrieval system or transmitted in any form or by any means, electronic, mechanical, photocopying, recording or otherwise, without prior permission in writing from Fonthill Media LLC

Typeset in 10pt on 13pt Sabon
Printed and bound in England

CONTENTS

Acknowledgments   5

Introduction   7

1   The Safest Coal Mine in the Country   19
2   A Tragedy That Never Should Have Happened   28
3   The Eight-Day Men: The Miracle in the Mine   88
4   Bringing Up the Bodies and Identifying the Victims   101
5   The Funerals   125
6   The Aftermath   162
7   The People   171
8   Revisiting the Tragedy   213
9   The Memorials   215
10   Cherry Today   220
11   The Victims   223

ACKNOWLEDGMENTS

Many thanks for information and pictures provided by John Taylor; Jack Rooney; Peggy Lami; Beth Bryant; Eric Sorebo; Michele Micetich of the Carbon Hill School Museum; DeAnn Pozzi of the Cherry Public Library and Museum; Fairbury Echoes Museum; Linda Willibey; and Mary Wren.

Thanks to descendants of the Cherry miners who provided new stories and pictures specifically for this book: Peggy Lami, Jack Rooney, Wendy Winbeckler, Stephanie Schmidt, Mike Scheiblein, Carmelita Collins, Alan Hatch, Mary Miller, Mike Kohr, Bob Steele, Lorena (Galletti) Cotton, Linda Sullivan, Anne Love-Hoskins, Joe Voyles, Susan (Matthews) Senders, and Jackie (Pierard) Uranich.

We all are indebted to research done over the years by Jack Rooney, Ray Tutaj, Jr., R. G. Bluemer, John Taylor, Michele Micetich, Richard Joyce, Edward Caldwell, Steve Stout, Karen Tintori, Jayson Tuntland, the Carbon Hill School Museum, the Cherry Public Library, the Coal City Public Library, and to all others who have accumulated a solid history of the Cherry mine disaster and coal mining in this region that will benefit generations to come.

I also want to give a special thanks to my wife, Janet, for her support of my work over the years.

I have wanted to write a book about the Cherry mine disaster for twenty years, but it was just out of my area. I am glad *America Through Time* has given me the opportunity. I have tried with this book, as I have with all my other books, to find material that has not been told before, adding to the general knowledge that has been provided by the state reports, local newspaper accounts, local historians, and the fine books listed here.

My intent with this book was to make this an illustrated history, to include more pictures of this event than have previously been seen by most people and to tell the stories of the people involved, which includes families of both the victims and the survivors.

Thank you to John Taylor for making available his large collection of coal mining picture postcards. Mr. Taylor earned a bachelor's degree in geology from Illinois State University and worked as a geologist for twenty-five years. He has written articles on mining and postcards and has contributed to several mineralogy books.

Recommended for further reading are the following books: *Fire Below* by R. G. Bluemer (2007); *Black Diamond Mines: A History of the Early Coal Mines of the Illinois River Valley* by R. G. Bluemer (2001); *The Cherry Mine Disaster* by F. P. Buck (1910); *The Promise of a Better Life* by Jim Piacenti (2011); *Trapped: The 1909 Cherry Mine Disaster* by Karen Tintori (2002); *The State of Illinois Bureau of Labor Statistics Report on The Cherry Mine Disaster* (1910); *From the Apennines to Spoon River: Stories of Migration from the Mountains of Bologna and Modena to America at the Turn of the Twentieth Century* by Pier Giorgio Ardeni; *Houses With Names: The Italian Immigrants of Highwood, Illinois* by Adria Bernardi; *Oneness Angiolina: The 1909 Cherry Mine Disaster* by Dean and Lorena (Galletti) Cotton; *Black Damp*, a novel about the Cherry Mine Disaster by Steve Stout (1979); *Cardiff: Ghost Town on the Prairie* by Jim Ridings (2006); and *Cardiff 2: A Second Volume of History from the Ghost Town on the Prairie* by Jim Ridings (2008).

Also used were newspaper articles from the *LaSalle Daily Tribune/News-Tribune* in 1909, 1910, 1977, 1979, and 2009; from the *Kankakee Daily Republican* in 1909; from the *Dwight Star & Herald* in 1909; and the books of the annual state coal mining reports. Also used were *Grande Disastro Della Mina Di Cherry*, the translation of the journal written in Italian by Antenore Quartaroli (1909) and *Eight Days in a Burning Mine*, by Thomas White, in *The World Wide Magazine*, October 1911.

About the Author

Jim Ridings was born in Joliet, Illinois, and graduated from Southern Illinois University with a degree in journalism. He won awards for investigative reporting and is the author of more than two dozen books of local history. Several of his books have won awards from the Illinois State Historical Society. He was a recipient of a Studs Terkel bronze medal from the Illinois Humanities Council in 2006.

INTRODUCTION

Coal mining has always been a dirty and dangerous job. Miners worked in dark, damp, and cramped conditions, breathing dirty or stale air. They did hard labor for eight or ten hours a day. There was always the danger of being injured or killed by falling rocks or coal, gas explosions, floods, fire, or a number of other dangers.

If accidents did not kill or maim them, lung disease did. In the early years of mining, there were no safety regulations or procedures required by the state. The mining companies did not always show the greatest concern for the safety of the workers. Replacements could always be found from among the immigrant workers pouring into America from Europe. Financial liability to injured or killed workmen was limited.

The miners did this hard work and lived this hard life on very low pay. Much of what a miner earned went back to the coal company to pay for what the miner needed to live, such as rent for their shack, or food and other items at the "company store," where the miners bought their supplies. A miner might not see his children except on Sundays. During the short winter days, some men never saw the sun except on Sundays.

The 1904 contract between the Illinois Coal Operators Association and the United Mine Workers of Illinois set the pay of miners at 61 cents per ton of coal mined. At some mines, the scales were rigged to cheat the miners.

Cherry miners were paid $1.08 per ton, and a two-man team could send up 5 tons a day. Cherry's mine was producing approximately 1,500 tons of coal a day—about 400,000 tons of coal a year in 1909. Thirty train cars each day hauled away the coal.

Workers such as tracklayers, drivers, water haulers, and other were paid from $2.23 to $2.42 per day. Sinkers (those who dug the mineshaft) were paid $2.87 per day. The contract also called for miners to be responsible for their own safety; miners whose work caused slate or rock to fall were required to repair the damage without pay.

When a man was killed in the mine, the contract called for each miner to contribute 50 cents for the man's family. The mine operator was to contribute $25. The miners were allowed to take off the rest of the day following a fatal accident, and they were allowed time off for the funeral. However, if the mine became idle because workers did not return to work between the time of the death and the time of the funeral, or if it became idle because workers did not return from the funeral quick enough, the mine owner did not have to pay the $25 to the victim's family.

Coal has been mined in seventy-three of Illinois' 102 counties, and more than 4,500 coal mines have been dug in Illinois since commercial mining began in 1810, according to the Illinois State Geological Survey (ISGS).

The ISGS estimates that coal is found under 37,000 square miles (68 percent) of the state. Most of the lower two-thirds of the state have coal underground. This amounts to more than 211 billion tons, or 112 billion tons in terms of minimum thickness. This coal reserve base is the second largest in the United States and is the largest in the nation for bituminous coal. The ISGS says Illinois coal resources hold more BTUs (British thermal units, a measure of heat) than all of Saudi Arabia's and Kuwait's oil reserves combined. The electric industry is the largest user of Illinois coal.

A total of 1.7 billion tons of coal were mined in Illinois' peak years of 1833 to 1924, according to the U.S. Department of Commerce, Bureau of Mines. Approximately 105,000 people were killed in coal mining accidents from 1900 to 2018, according to the U.S. Department of Labor, Mine Safety and Health Administration.

The Cherry mine disaster brought crowds of curious people. It also brought numerous photographers, and the scores of picture postcards they produced can still be found today in antique shops and on the internet. The majority of pictures in this book are picture postcard views taken at the time of the tragedy.

This coal auger drilled holes in the seam prior to blasting.

Introduction

This is an example of mining a wall, back-filling the right side with culm (coal and dirt waste from mining).

Mules were used to pull coal cars on rails from the mine on. They were stabled below ground.

The Illustrated History of the Cherry Mine Disaster of 1909

These coal-fueled boilers powered the hoisting equipment in the Cherry coal mine.

Introduction

The Cherry coal mine existed strictly to feed the engines of the Chicago, Milwaukee & St. Paul Railroad.

The St. Paul Coal Company office and the Cherry train depot. The office building later was moved to a site northeast of Princeton and was converted into a house.

Cherry historian Jack Rooney was told in the 1970s by Pete Donna (who worked in the Cherry mine at the time of the disaster) that miners, on their way home from work and covered in dirt, would conduct their business through a lower window near the back of the bank building. Next to the bank was the company store, Northern Mercantile.

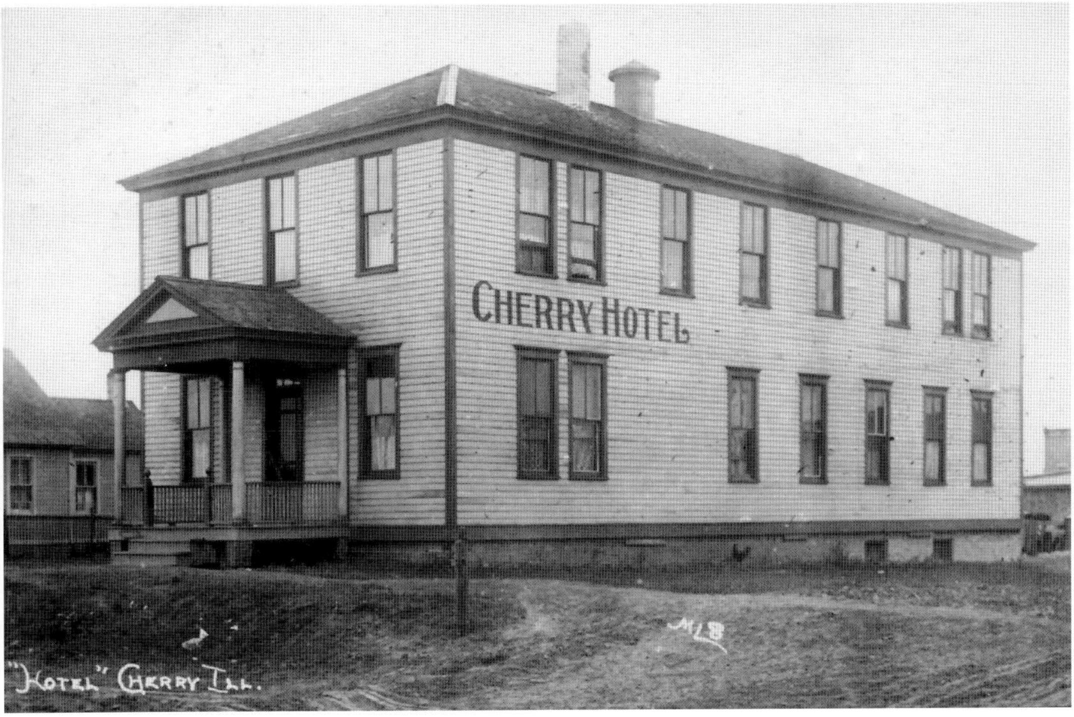
CHERRY HOTEL
"HOTEL" CHERRY ILL.
MLS

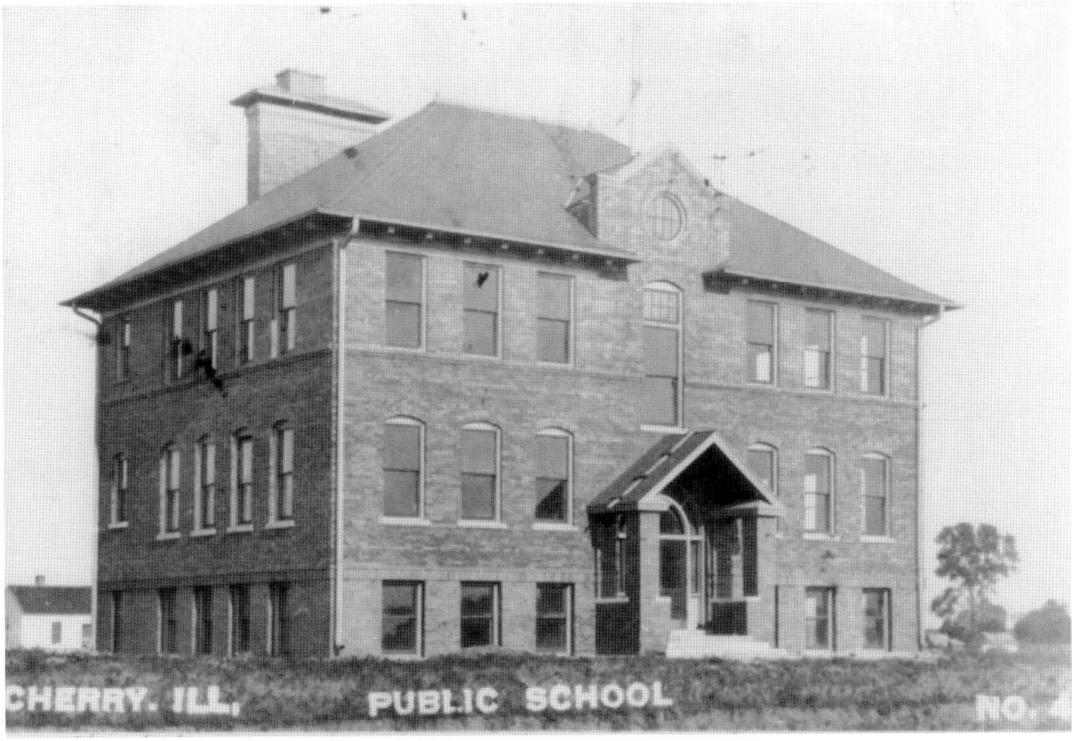
CHERRY. ILL. PUBLIC SCHOOL NO.

This unsigned postcard with a view of Cherry notes that the sender has worked two days. It was postmarked from Cherry on May 14, 1908, and sent to Mrs. Alf Balance in Cardiff, Illinois.

Below: F. C. Viner's store in Cherry.

Below: M. W. Ryan's grocery and hardware store and U.S. Post Office.

George Eddy is in the foreground of this picture inside an unidentified Cherry store.

Introduction

A baseball team from Cherry, pre-disaster. Tom Hewitt is on the left in the back row. William Hynds (who was killed in the mine) is on the left in the front row.

This Cherry baseball team in 1936 is from Holy Trinity Catholic Church's Holy Name Society. In front are Teo Rocco and Albert DeAngelo. Kneeling are Frank Reinsch, Dave Mazzetti, T. Giacomelli, Junior Donna, Mondo Ugolini, Albert Pacione, and Jim Giacomelli. Standing are Emmett Neill, Mike Jalley, Charles Bartoli, Matt Koster, Father Fred Winchell, Delio Joldate, John Jalley, and Leo Ferrarini. The team posed in front of the mine hills.

The Illustrated History of the Cherry Mine Disaster of 1909

Cherry school class, 1930s or 1940s.

Above left and middle: Here are two light-hearted pictures of Cherry men enjoying a drink. The barrel has the names of Camille Pieraud and his son, Jules, which may be a tribute to them for some reason, since they did not own a saloon. *Biere du Saison* is the name of a Belgian beer or pale ale.

Above right: Hector McAllister, a state mine inspector from Streator who was one of the men put in charge of operations at Cherry in the days following the mine fire.

1

THE SAFEST COAL MINE IN THE COUNTRY

Early on Saturday, November 13, 1909, approximately 481 coal miners entered the mine at Cherry, Illinois. By the middle of the day, a fire started that would doom 259 of them. By the end of the day, the mine was sealed in order to cut off the oxygen that was continuing to fuel the underground fire. The miners, both dead and alive, who were still in the mine at that time, were left down there, presumably for good.

The disaster at the coal mine at Cherry—officially called the St. Paul Coal Company Mine No. 2—remains the third worst coal mining disaster in American history. It is exceeded only by the disasters at Monongah, West Virginia, that killed 362 people in 1907, and the explosion at Dawson, New Mexico, that took 263 lives in 1913.

However, the mine at Cherry was considered the safest mine in the country, according to Warren Roberts, the engineer who built the mine. That statement was echoed by W. W. Taylor, St. Paul's general superintendent of mines. It had been in operation for just four years when the tragedy occurred.

The Cherry mine also was one of the largest coal mines in America. It was built of steel and stone, with modern conveniences such as electric lighting throughout the mine. The owners bragged that the mine was "fireproof."

Cherry was a typical coal town, a place that did not exist until the coal mine was established there. Coal was discovered there in 1904. The town was incorporated in 1905, the same year the St. Paul Coal Company (which was owned by the Chicago, Milwaukee & St. Paul Railroad) began hoisting coal there.

That area of Northern Illinois was rich in high-quality coal, and St. Paul Coal Company was establishing mines in a number of nearby locations. The town of Cherry was named for James Cherry, who was born in England in 1841. He started working in coal mines at the age of eight. He came to America when he was twenty-seven, working in a coal mine in Braidwood, Illinois. He worked his way up in the industry until he was named superintendent of the coal mine at Seatonville, where he was elected mayor. James Cherry became superintendent of the coal mines in Illinois for the St. Paul Coal Company, before being given the job as superintendent of the new mine at the place named for him. James and Elizabeth Cherry had eleven children. He died in 1909, two months before the mine disaster.

Cherry was a shipping mine, which meant its coal went to its railroad, not to the market for homes and businesses.

The St. Paul Coal Company not only spent a lot of money developing the new mine, it also spent a lot of money developing the new town that sprung up alongside the mine. It built a large grain elevator for the farmers in the area and it built a modern hotel. It built 125 company houses for its workers; another 250 private homes were built by new residents. There soon was a school, two churches, a bank, stores, and seventeen saloons, as well as whatever was needed for the instant town.

The coal was mined at three levels or veins. The first vein had poor-quality coal, so a second vein was dug at 320 feet. A few years later, a third vein was mined at 485 feet below ground.

James and Elizabeth Cherry.

The Safest Coal Mine in the Country

James Cherry, Jr., drives the last wagon load from the Wilmington Star No. 5 mine at Coal City, in 1927, as the mine closed.

The fan house and escape shaft on fire on November 13, 1909.

The Illustrated History of the Cherry Mine Disaster of 1909

Above left: Machinist and boss pumpman.

Above right: James Steele, mine superintendent.

The Safest Coal Mine in the Country

Mine managers at Cherry. George Eddy is in the back row in the center.

The Illustrated History of the Cherry Mine Disaster of 1909

The 3 o'clock Shift as they come from the Cherry Mine.

The Eleven o'clock Shift as they come from the Cherry Mine.

An unidentified miner in Cherry.

The Illustrated History of the Cherry Mine Disaster of 1909

CHERRY ILL.

The tipple at the Cherry mine. A tipple was where cars loaded with coal were tipped to unload them.

Unidentified mine workers.

On the left are two mine blacksmiths. Their kit is pictured above.

2

A TRAGEDY THAT NEVER SHOULD HAVE HAPPENED

The coal mine disaster at Cherry was a tragedy that easily could have been avoided. The fire was started by a careless error, and it got out of hand by a number of other mistakes, such as a failure to act quickly and decisively, a lack of adequate escape routes, miscommunication, and an inept handling of rescue efforts.

It began as a small fire—a common occurrence in the mines, which usually is quickly contained. That is why the miners were not alarmed at the first reports of the fire. They continued working, so they would not lose any time and pay from a needless evacuation. A few miners on their way out after their shift ended saw the small fire and they could have put it out, but they were not concerned enough. Coal continued to be dug and hoisted, on orders of the bosses, as the fire burned. The news of the fire traveled slowly to the miners until it was too late. It was forty-five minutes after the fire started before the general alarm went out to the miners who were still working.

The series of tragic events began about 12.30 p.m., when mine manager John Bundy ordered six bales of hay to the underground stables for the forty or more mules that pulled the coal cars along the rails. The electrical wiring for the lights in the mine was not working because of a short circuit, so kerosene torches were along the underground walls.

The hay in a coal car on the second vein was pushed by Robert Deans and Matt Francesco. The car sat under a kerosene torch. Kerosene dripped onto the hay, setting the hay on fire. Deans and Francesco reported the fire at about 1.25 p.m. to Alex Rosenjack, the hoisting engineer who operated the elevator cage between the levels of the veins. Rosenjack asked a few others to help him push the burning hay car toward the stables, where a water pump could be used to put out the fire. The car got stuck on the tracks. An exit door was opened, and the inflow of air fanned the fire.

Ray Tutaj, Jr., a local historian and Cherry mine authority, commented on his website:

> One of the most overlooked facts of the disaster, that no one ever talks about, is the reason the load of hay in the pit car was pushed around the runaround instead of the shortest route … It was simply because there were pieces of pipe and debris on the track, not too far from the main shaft. I read this in the Bureau County labor statistics report. It's just one little

A Tragedy That Never Should Have Happened

Smoke bellows as the fan house burns and smoke comes from the air shaft.

sentence that most people do not realize the significance of. I think there would have been no disaster if the workplace was cleaned up and made safe as soon as possible. The other fact was the electric lighting should have been replaced as soon as possible, but it wasn't.

The fire ignited timber supports. By the time the general alarm sounded to get out, the fire had burned into the coal seam, establishing itself and creating a deadly gas known as "black damp." The fire and the smoke prevented most of the miners from getting out.

The emergency whistle at the mine brought crowds of people from the town to aid in the rescue. George Eddy, a pit boss, came from home to the main shaft. He and Alex Norberg, another pit boss, organized efforts to stem the fire and get the miners out.

Norberg realized he needed to shut off the huge fan that blew fresh air into the tunnels because it was providing the oxygen that was fueling the fire. However, at the same time, it cut off the oxygen that the trapped men needed to breathe. Then, the fan blades were reversed to draw smoke out of the mine and prevent the fire from spreading to the mineshafts. However, that pulled the fire up the escape shaft, drawing the flames into the fan house, burning the fan and the fan house, and cutting off any possible escapes through the air shaft. Frank P. Buck, chief clerk at the mine, wrote the following in his 211-page book, *The Cherry Mine Disaster*, in 1910 about the tragedy:

> To the miner, the air shaft is the most vital thing in the mine ... Its value to his living is greater than his food. It means bringing into the damp, fetid and air-lacking depths, the God-given carbon and oxygen so essential to preservation.
> The interior of the mine became suddenly a volcano of fire, one vast fiery furnace, lit up till the light hurts men's eyes, almost as much as the torturing heat.

Workers tried to attach a hose to a water supply in the underground mule stables, but they were driven back by the heat and smoke. The only other water source was near the main shaft, but the hose did not fit the coupling. Screaming and moaning could be

A Tragedy That Never Should Have Happened

Anxious and worried family and townspeople hurried to the mine shaft when the alarm bell sounded.

heard from the miners trapped hundreds of feet below the surface and from the women waiting above ground. Buck wrote:

> Men that have never quavered before in life, whose lives were passed amid danger and suffering and who knew no fear, trembled with hideous terror ... It was the hearts of strong men shaken by the knowledge that other men were being tortured or about to be tortured, and that beyond the pale of assistance!
>
> [The flames] had become a mountain of roaring, gyrating, circumambient tongues of fire! The heat was unbearable. It burned their faces and even pierced through their rough mine clothing and scorched the skin underneath. It burned the hair and parched the lips until their voices died down to whispers.

Buck's prose continued:

> Men choked and gasped for breath and fell down and writhed on the ground as the death agony came upon them. The groans and shrieks and yells and screams of these poor unfortunate prisoners echoed like the cry of demons through the hewn corridors of the mine. Men grasped at men as they fell. The fingernails of many a man, as evidenced when his body was later found, tore into the soft flesh of his body when the death agony held him in its terrible grasp.

The rush to save those trapped became frantic. Twelve rescuers went down into the mine in the cage six times, bringing up miners who were choking and semi-conscious. Also brought up were a number of dead bodies.

It was the seventh trip underground that took the lives of twelve heroes.

John Cowley, the engineer of the main shaft, was ordered by Norberg to raise the cage only when the proper bell signal was given. The signals from the bell on this seventh trip were frantic and not the proper ones. So he did nothing. Miners on the surface screamed at Cowley to raise the cage, but he would not do it until they threatened his life. When the cage came to the surface, the people viewed the horrible scene of twelve men burned alive. Some of the men, and their clothing, were still on fire.

"When the cage reached the landing, the rescue party was aboard, not living men but corpses, seared and blackened, the tongues of flame having eaten into their flesh," *The LaSalle Daily Tribune* reported on November 15.

Cowley told *The Mendota Reporter,* "The men were all dead and their clothes smoldering when we got them to the top. Most of the bodies were grouped together. They were all standing up huddled together, grinning the grin of death itself."

The state report said:

> When the cage was raised, eight of them lay on the floor of the cage. Their clothing was still blazing and their arms and hands were in convulsive postures, just as death had seized them and when they had tried to protect their faces from the awful heat. Four of the bodies were lying across the top of the cage where they had died in a frantic effort to climb away from the fire.

"After the third trip and the subsequent announcement that the mine was to be re-sealed, cries and screams were heard from all parts of the crowd. The worst had come and now there was no chance for the fathers, sons and brothers below to escape," the LaSalle reporter wrote.

Norberg was one of the twelve men burned to death because Cowley followed the instructions he had given. The other men who gave their lives in the rescue efforts were John Bundy, mine manager; miners Andrew McLuckie, Harry Stewart, James Spiers, John Suhe, Robert Clark, and John Sczabrinski; driver, Joseph Robesa; Isaac "Ike" Lewis, a liveryman from Cherry; Dominic Foremento, a Cherry grocer; and John Flood, owner of a clothing store in town.

Sczabrinski was caught as the cage was lifted, and the top of his head was cut off. Buck wrote: "The sight was terrible, the blood had spurted out and sprinkled the clothing, faces and hands of the other eleven dead men. His clothing was still burning when the cage reached the top."

Herbert Lewis led the men who removed the dead bodies from the burning cage. He carried one victim a few feet away, put out the fire on the dead man's clothes, turned him over, and discovered it was his brother, Ike.

All but one of these twelve doomed rescuers were awarded the prestigious Carnegie Hero Award medal. John Suhe was not included because his father, Michael, was in the mine, and the Carnegie commission recognized only those who sacrificed their lives for those other than family members. Michael Suhe also died. George Eddy and Walter Waite, who miraculously survived in a separate heroic ordeal, also were awarded the Carnegie Hero medal.

There were many tales of heroism as the fire raged throughout the day. John McGill and his son were making their way through the smoke when the boy collapsed. McGill strapped the boy to his back with his belt and barely made it to a rescue team before collapsing. John Phillips dragged Edward Surrock along the tracks, both nearly running out of oxygen before making it out.

A Tragedy That Never Should Have Happened

John Flood.

Dominic Formento.

Above left: Dr. Lyston D. Howe and Herb Lewis.

Above middle and right: Isaac "Ike" Lewis and James Leadache, victims of the mine fire. Two others in his family were killed in the mine, Frank and Joseph Leadache, and two more were "Eight-Day Men" who survived after being buried alive for twenty-one days.

The Illustrated History of the Cherry Mine Disaster of 1909

Andrew McLuckie and the Carnegie medal he was awarded posthumously. The medal was donated by his family to the Cherry Library and Museum, where it is on display.

William Vickers and Steve Pohatney crawled on their hands and knees, over dead bodies, using rail tracks as a guide because they could not see, until they saw the faint lights of a hoisting cage. The bell signaled and the cage began to rise, taking up stricken miners. Vickers yelled, the cage stopped, and two miners made their way through the smoke to bring Vickers and Pohatney to the cage. Vickers later wrote:

> I escaped death by just three minutes ... When I arrived at the bottom of the shaft, the last cage was about to ascend. I shouted as the signal bell was ringing. Two men broke their way to me and dragged me to the cage. I then lost consciousness. When I came to, I was safely on top.

The newspapers were filled with other tales of bravery, including miners who gave their lives while saving others. The newspapers also were filled with tales of horror from men who survived and the agony of the women outside the mine. *The LaSalle Daily Tribune* reported:

> There has been weeping and wailing by the wives and children of those below, but they cannot comprehend the loss of life ... Wild-eyed women, with babes in their arms and tots trailing beside them, flocked around the mine, screaming, crying and yelling in their frenzy. All night long they crowded nearby, and again yesterday, in dry-eyed terror, they watched the work of the rescuers.
>
> Few are the homes in the village that have not lost one or more. One row of 48 houses was almost completely stripped of its wage-earning occupants; 46 of the 48 homes have

fathers or brothers missing. One old lady walked through the crowd yesterday, crying to herself. She had lost her husband and three sons.

Frank Buck wrote, as the cage raised the twelve burned men:

> The concerted cries of agony and despair were almost inhuman. So piercing were they and so filled with expressions of the deepest human emotions. Women threw their hands up to God and fainted by the scores on all sides. Hysteria reigned supreme in that multitude. Women writhed on the ground in agony, covering their faces and trying to shut out the appalling sight that was driving them insane. Others seemed to be bereft of reason and ran wildly about, shouting the names of their beloved at the top of their voices and crying out to know whether they had been saved or were dead. The crowd seemed to have suddenly lost their minds and were turned into an insane crowd, crying out for the victim that death had already claimed.

Rev. E. J. Ridings, pastor of LaSalle's Congregational Church, told the newspaper the women were "lost to reason," standing in the cold with light clothing, with their children who were turning blue from the cold.

One widow told the LaSalle newspaper that fourteen years earlier in the same month, her first husband was killed in a similar mine accident.

From Scotland, Mrs. Robert Love telegraphed to ask about her three sons. All were killed in the fire. One son had just written to his new bride to come to America; another son had just become engaged to a woman in Scotland.

Cherry mine officials ordered the mine be sealed at about 8 p.m. They believed anyone still below was dead, and they wanted to cut off the air to extinguish the fire. James Steele, mine superintendent, said that if the fire continued, it would collapse the

The Andrew Carnegie Hero medal was given to the men who died horrible deaths in a burning cage while rescuing miners: Alexander Norberg, John Bundy, Andrew McLuckie, Harry Stewart, James Spiers, Robert Clark, John Sczabrinski, Joseph Robesa, Isaac Lewis, Dominic Foremento, and John Flood, plus George Eddy and Walter Waite.

top works and disable the hoisting mechanism. It also was feared that air continuing to go down the airshaft would create a terrific explosion, such as happened at the mine in nearby Cardiff in 1903.

In explaining why the mine was sealed that night, Steele told the newspaper on November 16 that he believed every man in the mine was dead. He was asked how long a man could survive if all shafts of the mine were closed. "Not more than two hours," Steele replied.

The people on the surface, however, did not like the idea to seal the mine. They thought it meant doom for anyone who might still be alive, and they thought the company was doing this only to protect its property.

Rescue attempts continued in the following days, with a company from the Chicago Fire Department arriving on November 16. The situation was tense, so Governor Charles Deneen sent troops to maintain order on November 17. The governor put mine inspector Hector McAllister in charge of operations at Cherry on November 24.

On November 17, mine inspectors at the shaft told the crowd that everyone left in the mine was now dead. Mrs. Jerome Boucher screamed at them, "You let them die like dogs. You cowards, why don't you open the shaft and help the men! My husband is down in that hell, and you might have dug down to them by this time."

The newspaper reported, "She flung herself on the sand-covered shaft and lay there shouting imprecations on the experts until nurses carried her away."

Scenes like this were repeated over and over. When George McMullen's body was pulled from the mine on November 19, his widow "walked through the crowd with the trace of the smile of a half-crazed person, singing aloud, 'My George was killed in the mine, my George was killed in the mine.'"

Mrs. McMullen's mother, Mrs. Charles Dovin, died three days later. She had contracted pneumonia while waiting outside the mineshaft for several days in cold weather.

Andrew Dovin and his son, George, were among the dead.

In the book, *Houses with Names*, Josephine Fiore, who was a child at the time, said grieving women were trying to throw themselves into the mine to commit suicide. She said some people seemed to have intuition that morning, husbands and wives kissing each other goodbye when they never did before, and some not going to work at all.

Florence Eddy (1892–1975), the daughter of pit boss George Eddy, wrote about it for her Streator High School newspaper in 1910. The article is provided here by Alan Hatch, George Eddy's great-grandson. That day, she was told: "The mine in Cherry is on fire and your father is down below."

> In Cherry, everyone rushed to the mine, expecting of course, to have their loved ones safely up in a few minutes. Some women, whose husbands did not need to be at the mine, were comforting those whose husbands were below. Alas, too true, some of those very women who were thanking God that their husbands were safe found to their sorrow that they had gone down to aid those in the work.

She described when the cage rose, revealing the bodies of the men burned alive: "Cries of the children mingled with those of the wives and friends, rang in the air, and hearts almost stopped beating."

A Tragedy That Never Should Have Happened

Above left: This fine wooden statue of a coal miner stands in the Cherry Library today.

Above right: Unidentified mother and children.

"No one in Cherry thought, for a minute, of resting that night. Lights could be seen in every window, and especially the lights of the shaft overshadowed all others." She described sealing the mine, the men in helmets attempting rescues, and more. It is a well-written essay that adds much to the entire story.

By Thanksgiving Day, November 25, it was presumed that anyone still in the mine was dead. The shafts were sealed and not opened again until February 1.

The wire needed to fix the electric lights in the mine arrived on November 29.

A few days after the fire started, *The LaSalle Daily Tribune* reported a speech given to a crowd at the mineshaft by James Weatherbee: "I have been a miner for forty-six years, and I never in all my years have seen as big a blunder as has been made right on this spot. It was murder, that's what it was, and you'll all know it." He said he had witnessed mining disasters in England, Wales, West Virginia, and Pennsylvania and had seen mistakes in rescue work, "but never before have I witnessed such incompetency as you all see here."

The Illustrated History of the Cherry Mine Disaster of 1909

More than 25,000 curious people came by train and buggies to see the tragedy unfold. Special trains brought people from the Ladd station. There also was a large influx of photographers to record everything.

A Tragedy That Never Should Have Happened

C. M. & St. Paul Engine & Crowd enroute to Cherry at Ladd Ill. Nov. 15, 1909.
Photo by C. U. Williams, Bloomington, Ill.

Building Temporary Hoisting Apparatus, Nov. 15, 1909 Where 400 Men were Entombed Nov. 13, 1909 Mine Disaster at Cherry, Ill.

The Illustrated History of the Cherry Mine Disaster of 1909

A Tragedy That Never Should Have Happened

Women wait for news behind a roped-off area. State mine inspector Richard Newsam had the area around the mine shaft roped off to keep spectators back. He thought an explosion from the mine could be "a repeat of the Cardiff mine disaster." Explosions in the mine in nearby Cardiff killed nine men in 1903.

The Illustrated History of the Cherry Mine Disaster of 1909

The Chicago Fire Department helped tremendously by sending a lot of men and equipment to Cherry.

Cherry volunteer fire department. *Front from left*: Chief J.C Thompson, Assistant Chief Harley Daughtery, Mayor Hanney, Pete Donna, Bob McFadden, Andy Lettsome, Vic Gondoffi, Robert Maxwell, unidentified, Jimmy Kennan, Frank O'Brien, Thomas Hewitt, unidentified, Dick Cullen, and Billy Troy. *Back from left*: Tommy Crowe, unidentified, Fred Buck, Frank Mehoves, John Gunther, Bert Hanover, and James Billard.

A Tragedy That Never Should Have Happened

The Ladd Fire Department helped fight the mine fire.

The Illustrated History of the Cherry Mine Disaster of 1909

Plain boxes for coffins being unloaded from train cars.

A Tragedy That Never Should Have Happened

The Illustrated History of the Cherry Mine Disaster of 1909

The woman on the far left waiting with others to identify victims is Angiolina Sergenti.

A Tragedy That Never Should Have Happened

The Illustrated History of the Cherry Mine Disaster of 1909

A Tragedy That Never Should Have Happened

The Illustrated History of the Cherry Mine Disaster of 1909

Unidentified Cherry miners.

A Tragedy That Never Should Have Happened

The Illustrated History of the Cherry Mine Disaster of 1909

James Hand was the man in the helmet with the oxygen tank in an attempt to rescue men in the burning mine.

A Tragedy That Never Should Have Happened

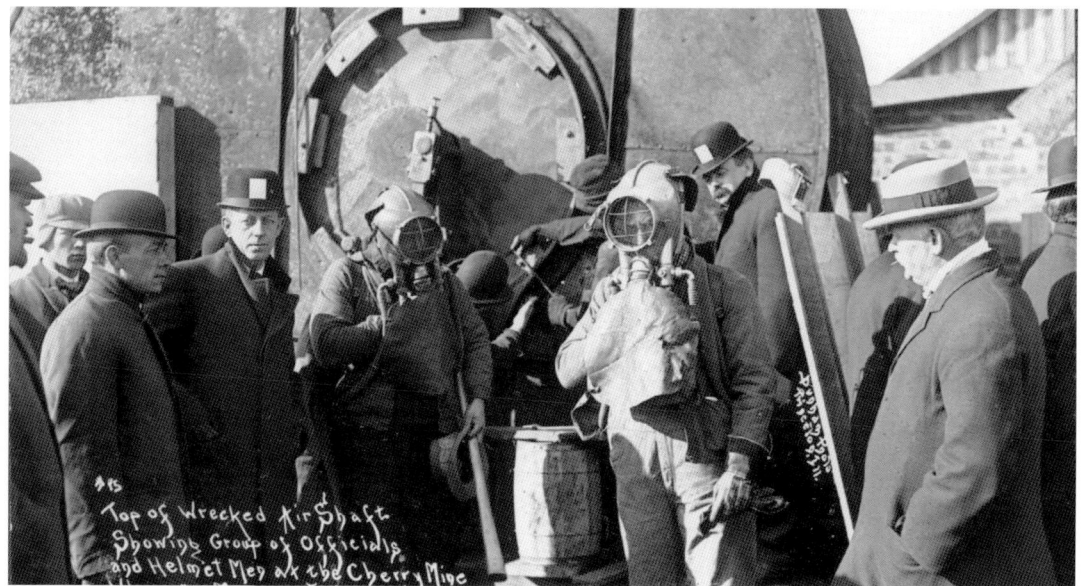

Above left: James Hand and Henry Smith volunteered to go down into the burning mine to rescue miners.

Above right: Henry Smith and R. Y. Williams put on safety gear to go down into the burning mine to rescue miners.

The Illustrated History of the Cherry Mine Disaster of 1909

A Tragedy That Never Should Have Happened

The Illustrated History of the Cherry Mine Disaster of 1909

No 18. Cherry Mine Disaster Showing shaft opened before fan was started Monday Nov. 15.

No 16 Cherry Mine Disaster Crowd Sunday Nov 14 09

A Tragedy That Never Should Have Happened

Barney Gately, machinist John Chedister, and Harry Dobbs pumped water from the mine's third vein.

This scene shows a cage being hosted from the main shaft, bringing up bodies.

The coal company's train brought needed supplies to the families of the victims.

Photographers from a wide area took numerous pictures, which were made into picture postcards.

No. 7 Cherry Mine Disaster Sunday Morning Crowd Viewing Mine from St. Paul Tracks

CHERRY ILL
Nov. 25-09

A.C. PERU ILL

97

The Illustrated History of the Cherry Mine Disaster of 1909

Trains brought thousands of spectators into Cherry in the days following the outbreak of the fire.

A Tragedy That Never Should Have Happened

Nuns from the Sisters of Mary of the Presentation and from St. Mary of Nazareth Hospital in Chicago kept families alive with money, food, clothing, child care, and help with household chores for the widows of the miners during the winter following the fire.

The Illustrated History of the Cherry Mine Disaster of 1909

Patrick Taggart, Jr., is the lead stretcher bearer carrying another victim from the mine.

A Tragedy That Never Should Have Happened

The Illustrated History of the Cherry Mine Disaster of 1909

A Tragedy That Never Should Have Happened

Ruins of the air shaft and fan house.

Cherry mine Superintendent James Steele is fourth from the right. W. W. Taylor, superintendent of mines for the St. Paul Coal Company, is fifth from the left.

The Illustrated History of the Cherry Mine Disaster of 1909

Cherry Mine Disaster Sunday Morning Crowd at the Escape Shaft. N°6

A Tragedy That Never Should Have Happened

The Illustrated History of the Cherry Mine Disaster of 1909

Above left: A mother comforts her children.

Above right: Observing the burned-out fan house and air shaft area.

A Tragedy That Never Should Have Happened

A photographer posed some of the miners who escaped with their lives. The clean-shaven man in the center is thought to be Frank Zanarini.

The Illustrated History of the Cherry Mine Disaster of 1909

Cherry Mine Disaster
Sunday Morning scene at the sealed shaft.   11-14-09

A Tragedy That Never Should Have Happened

Taking up the first body after reopening the mine on February 4.

The Illustrated History of the Cherry Mine Disaster of 1909

Mine inspectors from several states came to Cherry to investigate this major event.

A Tragedy That Never Should Have Happened

National Guard troops from Kewanee and Galesburg arrived to keep order.

73

The Illustrated History of the Cherry Mine Disaster of 1909

A Tragedy That Never Should Have Happened

Illinois National Guard soldiers on duty in Cherry.

The Illustrated History of the Cherry Mine Disaster of 1909

Sixth Regiment, Company "K" of the Illinois National Guard, out of Kewanee, was under the command of Captain W. F. Hall.

A Tragedy That Never Should Have Happened

Rail pit cars brought up from the mine after the fire.

The Illustrated History of the Cherry Mine Disaster of 1909

An overturned pit car.

Scene at Mine Disaster, Nov. 13, 1909 Cherry, Ill.

A Tragedy That Never Should Have Happened

No. 15 Cherry Mine Disaster
Fan House + Escape Shaft Sunday P.M. Nov. 14th

The Illustrated History of the Cherry Mine Disaster of 1909

A Tragedy That Never Should Have Happened

Scene at Mine Disaster Showing Wrecked Air Shaft Nov. 13, 1909, Cherry, Ill.

Cherry mine disaster Nov. 13, '09.

The Illustrated History of the Cherry Mine Disaster of 1909

A Tragedy That Never Should Have Happened

Knights of Pythias lodge members in Cherry.

Above right: W. W. Taylor, general superintendent of mines for the St. Paul Coal Company.

The Illustrated History of the Cherry Mine Disaster of 1909

W. M. MacClain stands in between Tom White and John Lorimer, two of the men who survived for eight days underground. MacClain was editor of *Appeal to Reason*, a socialist newspaper in Kansas.

A Tragedy That Never Should Have Happened

Waiting for news by the mine shaft.

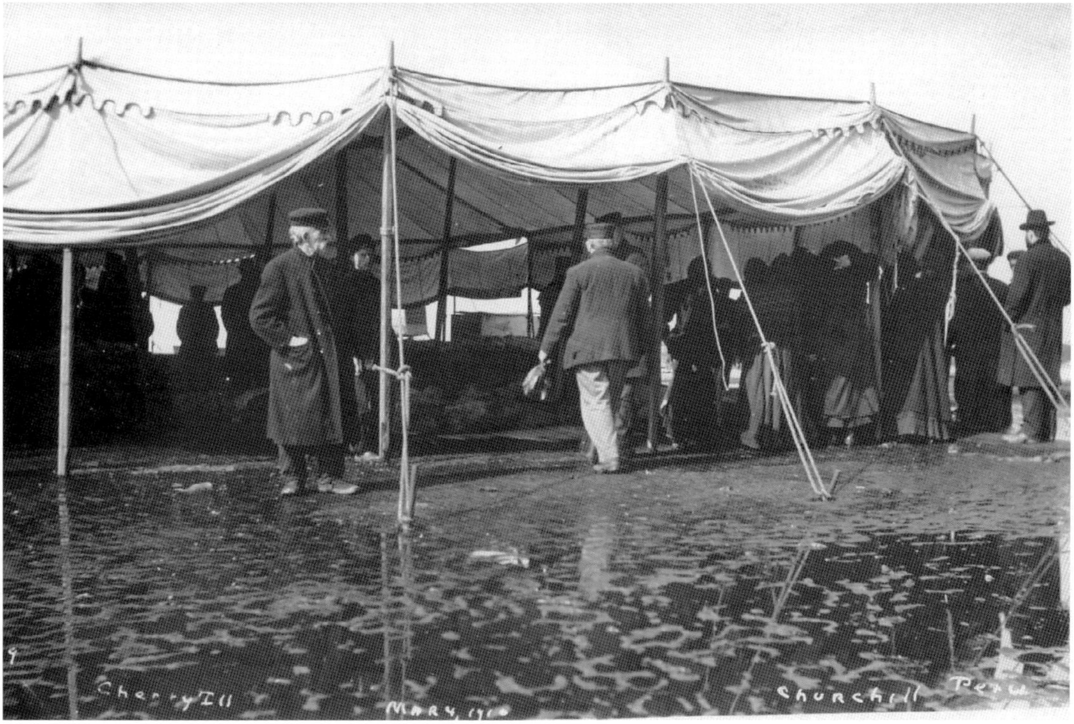

Another grim day at the morgue tent.

A train leaves Cherry on March 6, 1910, with coffins for funerals in other towns.

Above: A scene outside the mine opening while the fire roared.

Below: Inside the Cherry mine—Elmer Lettsome (father-in-law of Eliza Parker of Cherry and Cardiff) is in the front on the right, holding his lunch pail. Behind him with a pipe is Mr. Tintori.

A Tragedy That Never Should Have Happened

Samuel Howard.   Walter Waite.   Dr. Lyston D. Howe.

Jack Railsbeck, a mechanic and cage engineer; Frank Buck, company clerk; L. C. Kingery, a writer; and Walter Waite, assistant mine manager who survived eight days underground.

3

THE EIGHT DAY MEN: THE MIRACLE IN THE MINE

The mine was opened for recovery efforts, and then closed because it was too dangerous, several times during the week following the outbreak of the fire. Eight days after the tragedy, the mine was unsealed again. It was thought the fire would be out by then, so the recovery of bodies could begin. However, a miracle happened. Twenty-one men were found alive. They had survived by drinking what little water seeped into the darkness and by chewing their hatbands, shoes, belts, elm bark from the props, old lemon peels, and even the "sunshine" wax from their head lamps.

The miracle survivors were George Eddy, Walter Waite, Frank Waite, Thomas White, John Lorimer, John Thomas Brown, John Barnoski, John Semich, George Stimac, Frank Zanarini, Antenore Quartaroli, Daniel Holafcak, William Clelland, Fred Lauzi, Salvatore Pigatti, Giacomo "Joe" Pigatti, Bonfiglio Ruggeri, Fred Prohaska, Frank Prohaska, John Lemache, and Georgio Lemache.

It was Eddy, a member of the rescue crew, who found himself trapped and led the men to the area in the mine where they could survive.

On Sunday, the day after the disaster started, Walter Waite suggested that each of the men write a letter to their families and put it in their pockets. If they died, the letters would be found on their bodies. William Clelland had a notebook and a pencil. Each man wrote his note.

Giacomo "Joe" Pigati wrote:

> This is the fourth day that we have been down here. That's what I think, but our watches stopped. I am writing in the dark because we have been eating wax from our safety lamps. I also have eaten a plug of tobacco, some bark, and some of my shoe. I could only chew it. I hope you can read this. I am not afraid to die, Oh, Holy Virgin, have mercy on me. I think my time has come. You know what my property is. We worked for it together and it is all yours. This is my will and you must keep it. You have been a good wife. May the Holy Virgin guard you. It has been very quiet down here, and I wonder what has become of my comrades. Goodbye until heaven shall bring us together.

Walter Waite wrote:

The Eight-Day Men: The Miracle in the Mine

Dear Wife and Family, I write these few lines to let you know that we was alive at this time and if we are found dead, try and keep the family together and use your best judgment about what you can do for them, for I may not see any of you again, so good bye and God Bless you all. From your Loving Husband and Father, Walter Waite.

Walter Waite's brother, Charles, was one of the men killed in the mine.
John Lorimer wrote:

Dear Wife, I am still living yet this is five o'clock Sunday morning, but we have poor hope, as the blackdamp is getting the best of us. There is twenty-one of us all together here, dear wife. Don't grieve, we will meet again. God bless you and believe in Him. He will take care of you. I guess we will meet in a better land, when you get over this let them know at home, that is all, Dear, God bless you. Your loving husband.

Antenore Quartaroli's note to his wife said:

Dear Erminia and Son: Now being half past one p.m. I am very hungry but suffering much more from thirst. Dear Erminia, am very sure that my last hour has struck, and never will leave this grave, I beg of you not to think no longer of my death for I feel I will have an easy death. You will write to my unfortunate mother and brothers and tell them of my sad death. I have nothing more to say, only that to educate my dear child the best you can, and

Antenore Quartaroli.   Joe Pigati.   George Eddy.

when he grows you may tell him that he had an honest father, would like to say hoping to see you again, but must say goodbye forever, last kisses from your Antenore.

George Eddy wrote:

Dear Wife and Children. I write these few lines to you and I think it will be for the last time. I have tried to get out twice but was driven back, there seems to be no hope for us, I come down this shaft yesterday to help save men's lives. I hope the men I got out were saved. Well, Elizabeth, if I am found dead take me to Streator to bury and move back. Keep Esther and Florence and Jennie together as much as you can. I hope they will not forget their Father so I will bid you all goodbye and God Bless you all.

Tom White wrote:

Dear Wife and Children, I am now writing just before we all go. I know, Maggie, you will be in a awful state. I have been thinking of you, Mag, and the children. I loved my children and wife. But if it is God's wish for us to go, God knows what is best. It is five o'clock Sunday morning when I am writing. Maggie, I am praying to God and my Savior. Good-by wife and children. Be good to the children, Maggie. Please give all the folks at home my best wishes. Maggie, I wish you and the children to attend church and live good Christian lives, believe in God, Maggie. From your Loving Husband. Tom White.

Antenore Quartaroli wrote a much longer account after his rescue. He was working when a pit boss came by at 2 p.m. to measure his work. They talked for a while, and although the boss knew about the fire, he did not mention a word. Quartaroli smelled smoke at 2.55 p.m., then saw a cloud of smoke coming toward him rapidly.

Quartaroli alerted a fellow miner and they grabbed their coats and lunch buckets, then ran toward the escape shaft, but they were stopped by the smoke and heat. Quartaroli told his friend, "Well, Frank, here's where we came to die, and I'm sure this is the last moment of our life." His friend told him to not give up.

His lengthy story told of his efforts to survive the heat and smoke, and how he and other men found an area in the mine to take refuge after the reversal of the fan sent deadly smoke towards them. They talked, asking one another how the fire started. Eddy told them it started from a bale of hay. "None of us could believe that one bale of hay in flames could cause so much tragedy," Quartaroli wrote.

The twenty-one men tried to find a way out, but every route was blocked. "I heard a voice say, 'Even here there is black damp (methane gas). We can't go ahead.' At hearing this, a cry of terror went from mouth to mouth, and if anyone had seen us at that moment, they would not have known us as human beings," Quartaroli wrote.

They looked for a safe place to take refuge. "My pen is not strong enough to describe the fatigue I suffered," Quartaroli wrote. "I was so weak, hungry and thirsty that I could barely lift my legs, and I said to myself, 'How can a man at my age be reduced to this state?'"

Quartaroli collapsed and was helped to his feet by William Clelland, and they marched on. They found a place with better air. Hungry, they contemplated killing a mule, but the black damp prevented them from reaching the stables, and they believed they would be rescued by morning. They found some lunchboxes abandoned by other miners, but all they got was one slice of bread and half a bottle of water.

George Eddy lay on the ground and cried. He said, "We have slaved all of our life and sacrificed to raise our families, and God has recompensed us with this death." Quartaroli wrote:

> Eddy was becoming delirious and mentioned his wife, children and many things this pen can't write. Hearing Eddy in his delirium made the rest of us become thoughtful, not of the fear of death, because as yet we did not believe we would die so young. But Eddy's words reminded me of my wife, child, brothers and my mother. I thought of what would happen if I didn't see them any more, and if I die, who would take care of them, and many other things.

The account Quartaroli wrote after his rescue included information about the terrible thirst, hunger, and fatigue he suffered. He had many stories of desperation and heroism. A Belgian named Leopold Dumont (nicknamed Paolo, and called a Frenchman by the others) tried to find his way out on his own, despite warnings from the others. Black damp overcame him. The other men tried to revive him, but Dumont died.

Walter Waite spoke. "Now there is no hope of departing alive from this tomb. We must resign ourselves to die as men. Keep carefully the few lines you wrote. Before dying, however, it is my idea to pray to God not to give us a cruel death." Waite led the men in prayer.

At 9 p.m. Sunday, William Clelland said, "Brothers, if you can try to sleep, do so. In that way, the black damp will send you to the other world quicker, without suffering."

The men tried to sleep. Quartaroli wrote:

> After a while, Walter Waite arose and squeezed our hands, and at that moment we could only say Goodbye Forever. In that position, I spent an hour of terrible delirium, and I'll never forget that hour in which 20 men, whole and strong and of generous heart, lay on the ground near each other waiting for eternal sleep. I shall never forget it if I live a thousand years.

A major factor in helping the men survive was Waite's suggestion that they build a wall to seal their area from the smoke and the black damp. The men were almost too weak to do it, but they managed. It bought them more time, and it saved their lives.

Quartaroli's compelling story would make a great movie. There is a lot of detail about bravery and survival, and even treachery by one man who found water dripping and would not tell the others, right up until the eighth day when they were rescued and reunited with their loved ones. One of the miners died the day after being rescued. Daniel Holafcak was fifty years old and suffered from asthma, and did not fare well throughout his ordeal.

The Des Moines News reported:

> One man asked only to see his wife and children, he did not want food, he said. Another begged for a glass of beer, while a third scorned food but remarked, "Lord, how I wish I had a cigarette."
>
> Despite their affected jauntiness when they were rescued, the faces of most of the survivors plainly tell the story of their sufferings. A week ago William Clelland's hair was a dark brown; tonight it is a silver gray. He was almost too weak to walk until a childish voice called his name through the window of the sleeping car, when he was strong enough

to reach out and gather his two children into his arms. His six year old son Willie, and his eight year old daughter Frances, perched themselves upon his knees and the first thing Willie said was, "Papa did you get your dinner?"

The newspaper account continued:

> "It was the greatest miracle of the age," declared William Taylor, state mine inspector, as the tears rolled down his cheeks. "This is the greatest moment of my life. No one could feel the enormity of the disaster more than I, and that men are really alive down there is nothing but a direct answer to our prayers. Nobody can make me believe otherwise."
>
> On all sides tonight, men are saying, "Any man who does not believe in God after this is simply insane."

John Brown's wife, Mary, did not give up hope that her husband would be found alive. Mine officials immediately offered to pay the fare for Mary and her children to return to England, from where they had come just two months earlier. She refused. After his rescue, John had offers to tour the country to speak about his ordeal. Instead, he and his family moved to Des Moines, Iowa, where he worked in another coal mine until 1934, and then he went into construction work.

Several other miners who did not survive wrote notes, found on their bodies later. Samuel Howard wrote:

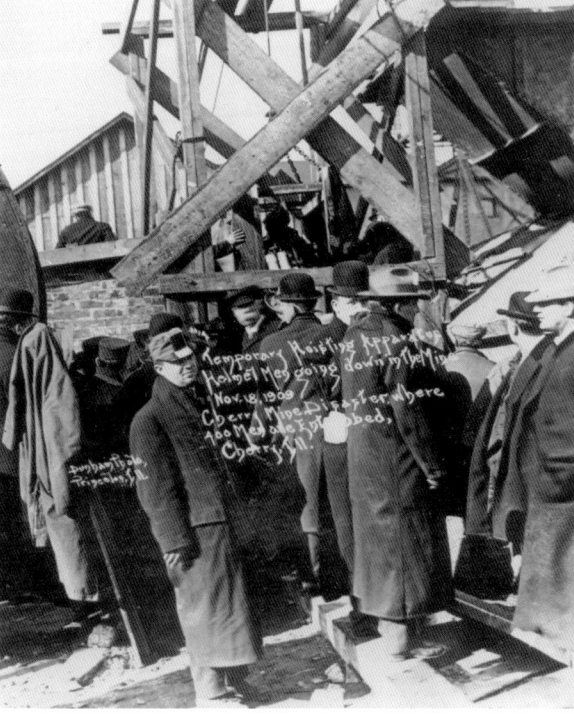

Above left: George Eddy's note to his wife from inside the burning Cherry mine.

3:25 p.m. The black damp is coming on us. Only for the (hand-made) fans we would be dead. Dying for the want of air. 6:20 p.m. Now we are trying to make the bottom with our fans. We have six of them moving. 9:15 p.m. Sunday. Still alive. We have to come back with our fans. 10:25 p.m. Sunday. Eventually we all had to come back. We can't move forward or back. We can stand it with our fans until Monday morning. Am still alive. We are cold, hungry, weak, sick and everything else.

His last writing was on Monday afternoon: "The lives are going out. I think this is our last."

Sam and his fourteen-year-old brother, Alfred (birth name Develeschoward), died together in the mine. Sam Howard was engaged to Mamie Robinson and had just purchased a diamond ring that he never had the chance to give her. As he lay dying in the mine, he wrote a note saying that the ring was at the post office, and if he died, the ring was to be given to Mamie. It was done, and she wore the ring for a long time, before eventually getting married and raising a family in nearby Arlington.

A few other miners missed their miracle by only a few hours. It was reported on November 21—a day after the twenty-one miracle miners were saved—that the bodies of thirty-seven miners were found. They had been dead just a short time. If rescuers had arrived just a few hours earlier, the newspaper reported, they might have been saved.

The covers of two Italian magazines, with stories on the men rescued after eight days.

The Illustrated History of the Cherry Mine Disaster of 1909

Two illustrations depicting the men's ordeal in the mine, from *Wide World Magazine*, October 1911.

Illustrations from *Saga* magazine, September 1953, for a story about the men trapped for eight days.

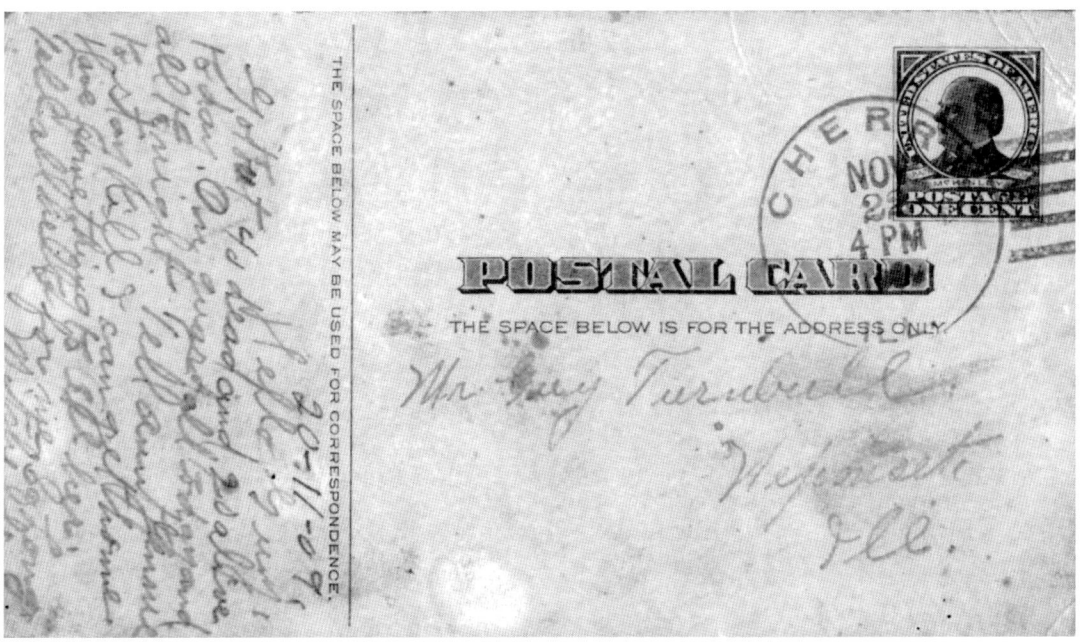

Postcard dated November 20, postmarked November 22 from Cherry, from a National Guardsman to Gary Turnbull in Neponset, Illinois: "Got out 40 dead and 20 alive today. On guard all day and all night. Tell Aunt Jennie to stay til I can get home. Have something to tell her. Tell all hello for me. So long. Mitchell."

The fan blades on which a miner wrote, "All alive—2 p.m., 14." The bodies of that group of thirty-one miners were found on April 10, 1910.

The Illustrated History of the Cherry Mine Disaster of 1909

A postcard sent by Domenica Compasso, widow of victim John Compasso, to her sister, Rose (Mrs. Giovanni) Gaudio, in Joliet, Illinois. She writes in Italian that her nephew, Johnny, was staying with her in Cherry and will be going to Morris and not to Joliet. The picture shows the fan made by the men trapped in the mine.

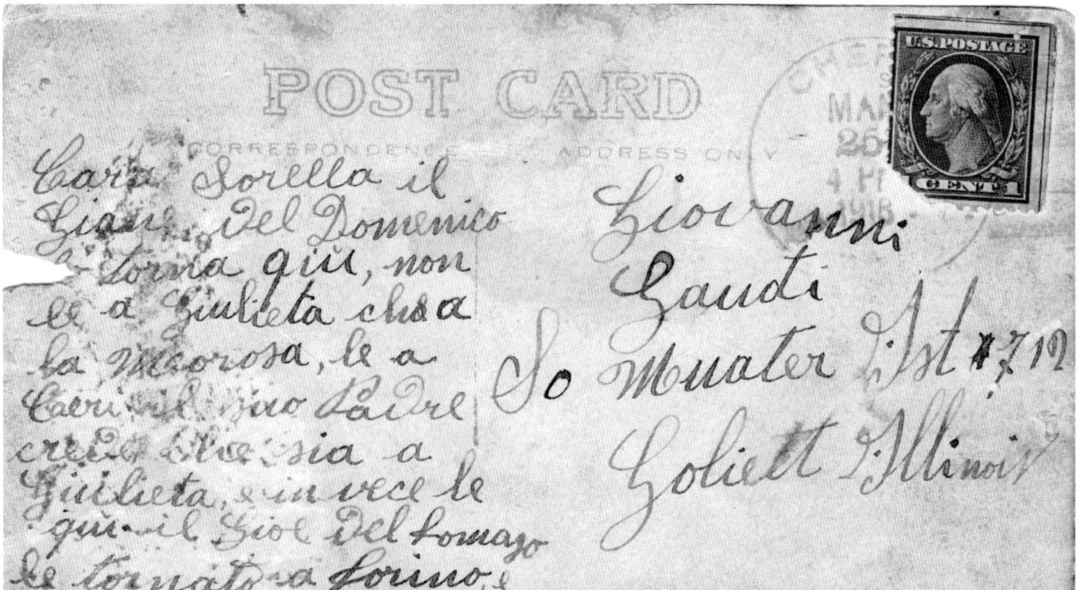

The Eight-Day Men: The Miracle in the Mine

Shouts of "They are alive!" came from the crowd on November 20. A group of twenty-one men were found alive underground, eight days after the fire started.

Some of the miners who did heroic work rescuing their fellow men.

The Illustrated History of the Cherry Mine Disaster of 1909

Mine rescue team from Standard, Illinois.

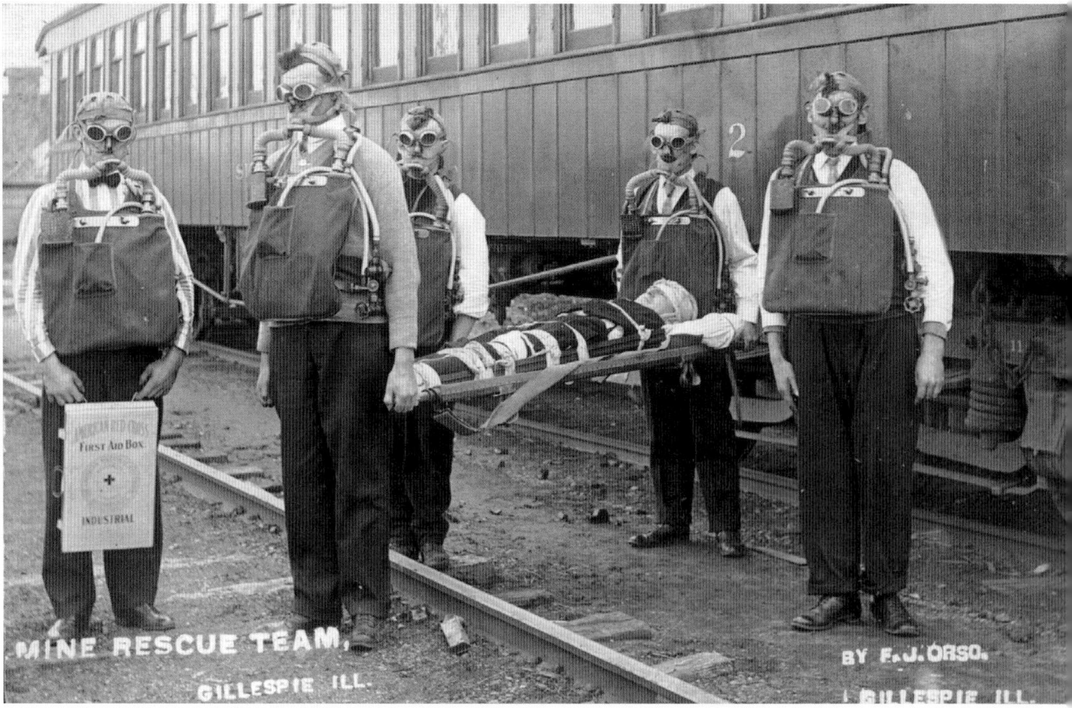

The Eight-Day Men: The Miracle in the Mine

Special train sleeping cars were used for the "Eight-Day Men" to recover after being rescued on November 20.

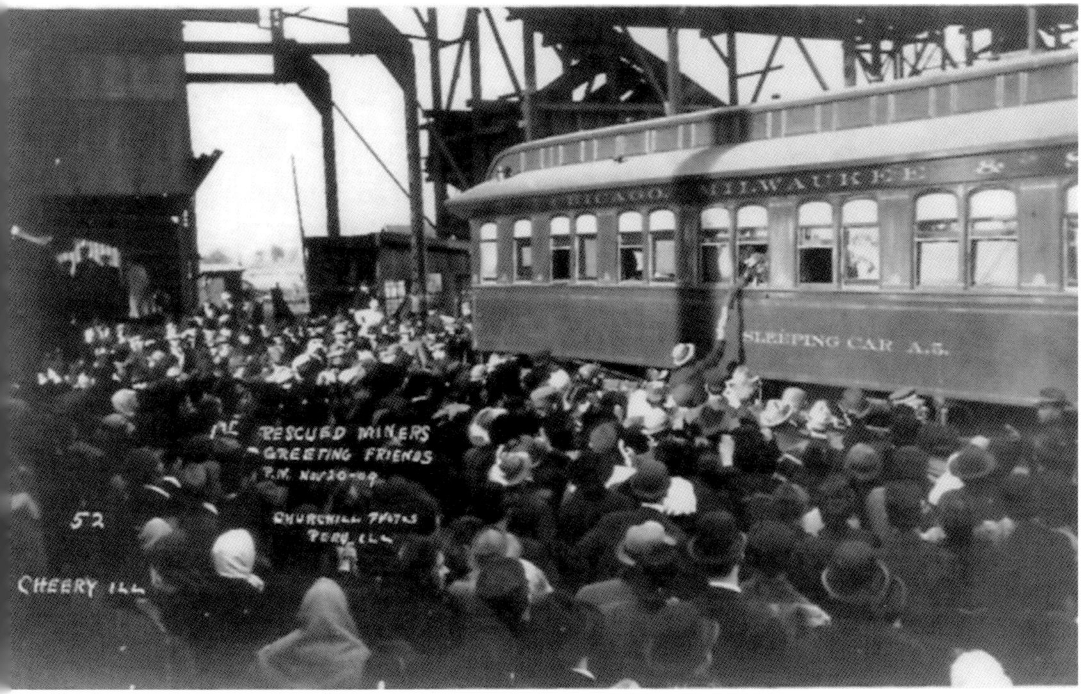

The Illustrated History of the Cherry Mine Disaster of 1909

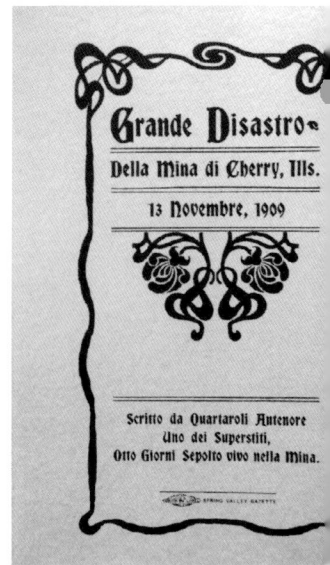

Above left: George Eddy, his wife, and daughter, with mine superintendent James Steele on the right.

Above right: The cover of Antenore Quartaroli's amazing account of how he and the others survived eight days in the mine.

Undertakers at the scene on February 20 were, from left, M. Knauff, W. Kingery, J. Bailey, J. McCann, and three helpers.

4

BRINGING UP THE BODIES AND IDENTIFYING THE VICTIMS

The first body to be recovered was Ole Frieberg on November 18. The mine was sealed again on November 25.

As would be expected, the disaster at Cherry was big news all over the world. Firefighters from Cherry and Ladd responded quickly. The Chicago Fire Department sent men and equipment. The Red Cross, the Knights of Pythias and various churches sent aid. The *Chicago Tribune* sent medical personnel and started a relief fund. Most of the miners were immigrants from across Europe, and foreign governments sent delegations to look after their countrymen. The Italian government offered to pay passage for any miners and their families who wished to return to Italy. Newspapers all over the nation appealed to their readers to send aid.

The mine was unsealed on February 1, 1910, in order to retrieve the bodies of the dead. Passages were cleared before the bodies could be reached. The fire was still smoldering in the mine after all those months.

Thirty-one men died together in one corner of the mine. In their last hours, they built two fans out of 12-inch boards, measuring 4 feet in diameter. On one blade was written, "All Alive, 2 p.m., 14." That was the day after the fire started. The fan and the bodies of the men were discovered on April 10, 1910. The fan was crated and was to be put on a train to Chicago. A large crowd gathered at the train station. Overcome with emotion, the crowd (mostly women) attacked the train people, the mine people, the sheriff, and his deputies. They tore open the crate and took the fans, leading a parade through Cherry and causing a near riot. The fans were kept in the homes of two widows.

Some bodies disintegrated when they were brought up into the fresh air after all that time underground. Family members saw flesh, heads, and limbs fall off their loved ones. Some victims were so decomposed that they could not be identified, not even by family members. Dozens of bodies became mummified after months underground.

The last of the group of bodies was found on April 15, 1910, in the second vein. Another victim was found on July 7, buried under a roof fall. The last two were found on September 11; they were unidentifiable, but it eventually was determined they were brothers Frank and Lewis Bawman.

The Illustrated History of the Cherry Mine Disaster of 1909

These first four bodies in the morgue tent were the Love brothers: John, Morrison, David, and James.

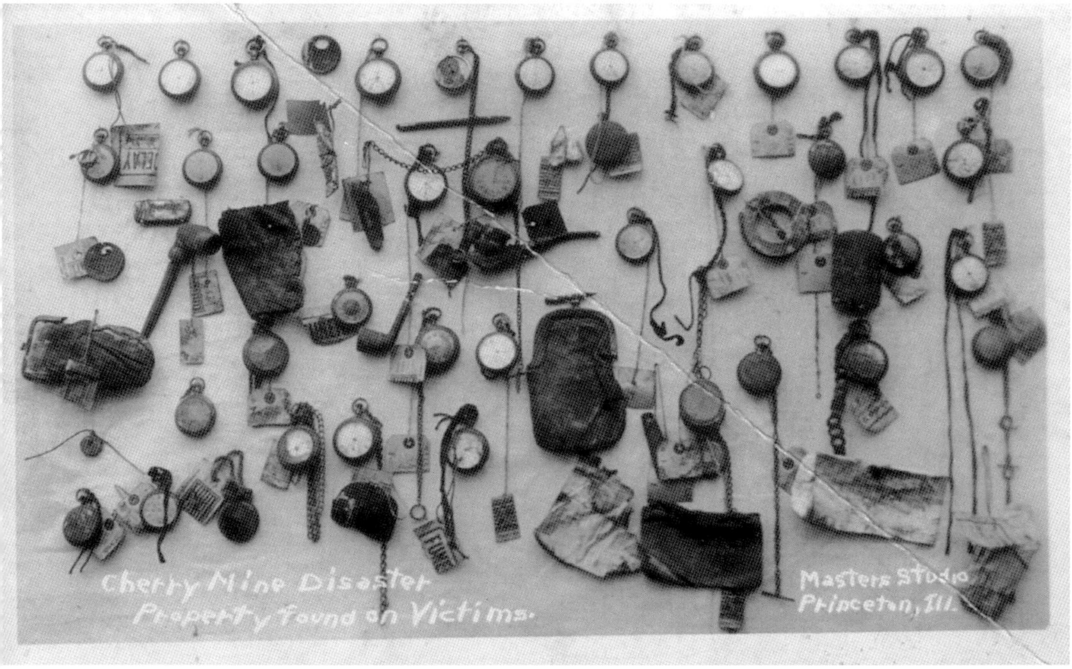

Bringing Up the Bodies and Identifying the Victims

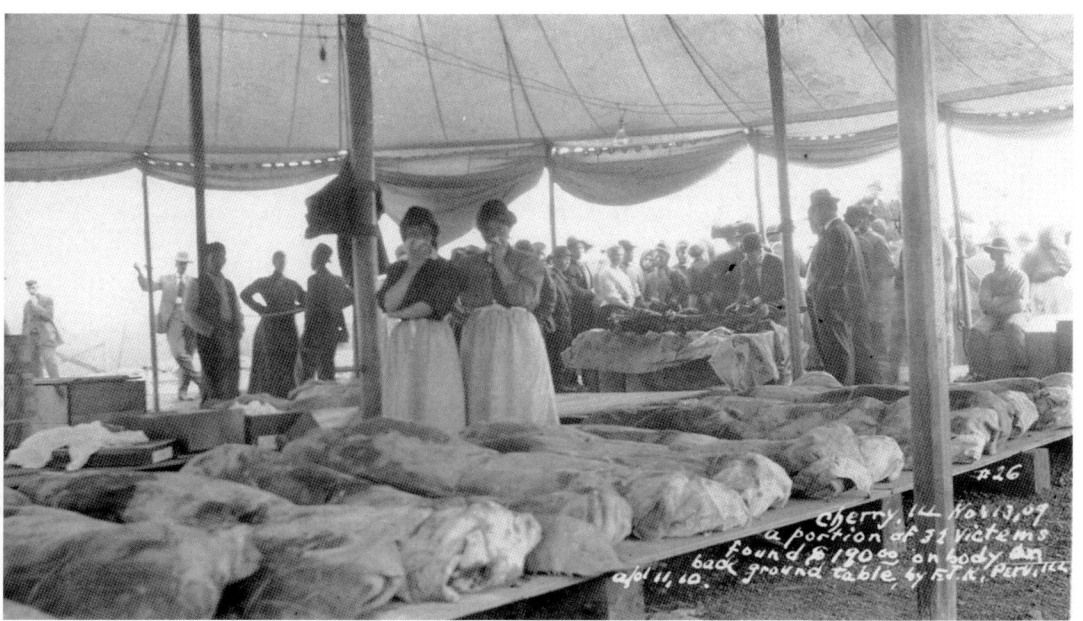

A morgue tent was set up. After several months in the mine, most of the bodies could only be identified by personal possession or by their numbered tag. Bodies were identified and funerals were held for months. St. Paul Mine donated a piece of land for a cemetery. In some cases, a trench was dug instead of individual graves, and the caskets were buried side by side. A number of miners were buried in other cemeteries in other towns.

The Illustrated History of the Cherry Mine Disaster of 1909

Peter Donna was sixteen years old in 1909. He was in the third vein of the Cherry mine, shoveling the coal that was loosened by the pick of his father, John Donna. They saw the smoke and went up to the second level. Fire blocked them from going further; however, they eventually made their way to the main elevator cage, with their hair singed and coughing black smoke. Peter—who died on September 19, 1977, in St. Margaret's Hospital in Spring Valley, Illinois, at the age of eighty-three—later wrote about the event and gave talks up until the time he died. He repeated that the Cherry disaster was "The biggest bunch of carelessness I have ever seen."

Cherry historian Ray Tutaj, Jr., agrees. "A heap of human carelessness led to the disaster. It should have never happened."

The state report issued in 1910 delicately put it, "Associated with all great calamities are some simple, curious or mysterious causes."

Bringing Up the Bodies and Identifying the Victims

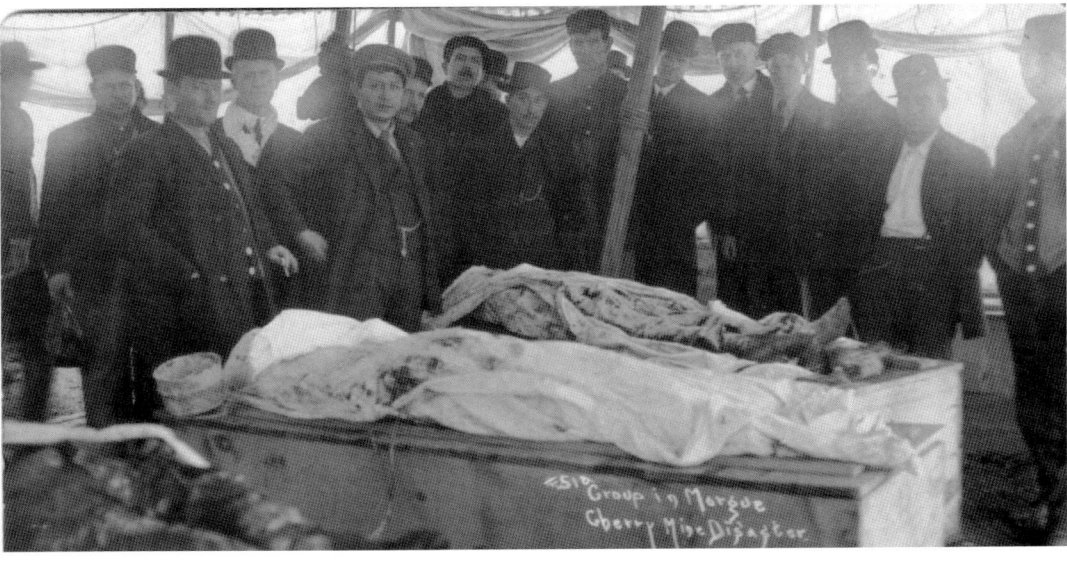

The Illustrated History of the Cherry Mine Disaster of 1909

Searching the body of a victim inside the morgue tent.

Bringing Up the Bodies and Identifying the Victims

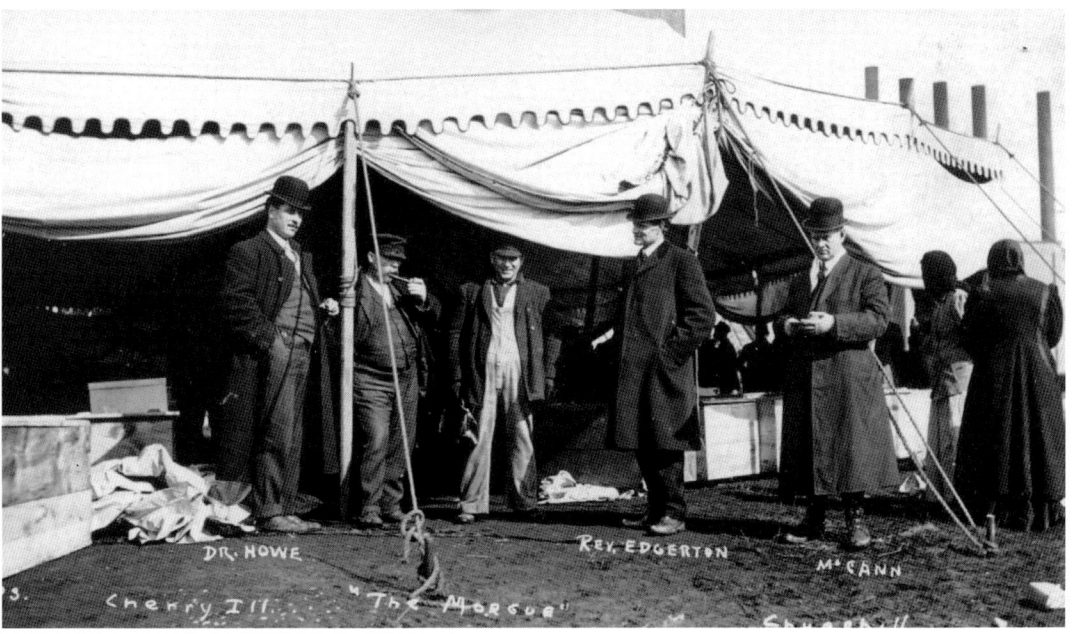

A train car unloads caskets for the mine victims.

Dr. Howe, Rev. Edgerton, and Mr. McCann outside the morgue tent.

The Illustrated History of the Cherry Mine Disaster of 1909

Bodies are put in caskets and hauled away from the morgue tent.

Bringing Up the Bodies and Identifying the Victims

Bodies were placed in boxes as they were brought up and identified.

111

The Illustrated History of the Cherry Mine Disaster of 1909

#21 Known by check as Body of Anton Krovonia, found $200⁰⁰ on his body, recovered from 3rd level apl 11. 1910. FJK photo Peru. Ill. cherry. Ill. Nov 13. 09

These two pictures show the body of Anton Krovonia, identified by a check number on his remains. He had $200 in his pocket.

#10 Sisters of charity consoling grief stricken Relative of Anton Krovonia, recovered 5 months after cherry mine accident with $200⁰⁰ on his person, cherry. Ill. Nov 13. 09 april 11. 1910. FJK photo Peru. Ill.

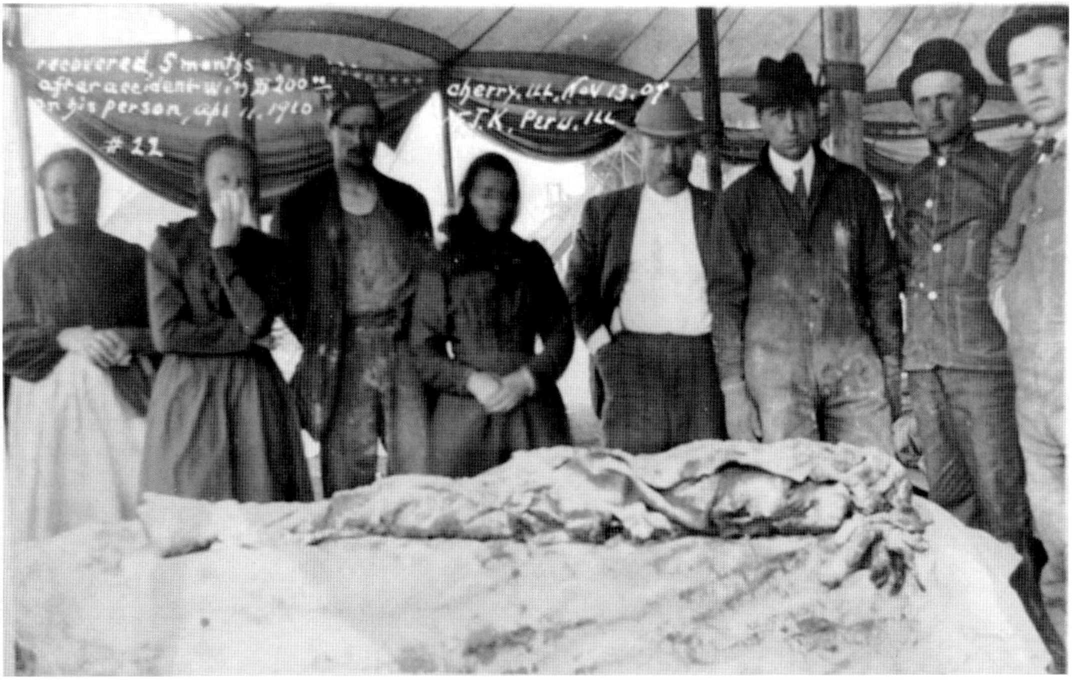

Bringing Up the Bodies and Identifying the Victims

This is what was left of Anton Krovonia's body, with the $200 in his pocket. That was a tremendous amount of cash for a miner to have at that time, but many miners did not trust banks.

A quadruple funeral procession begins at the mine tipple.

The Illustrated History of the Cherry Mine Disaster of 1909

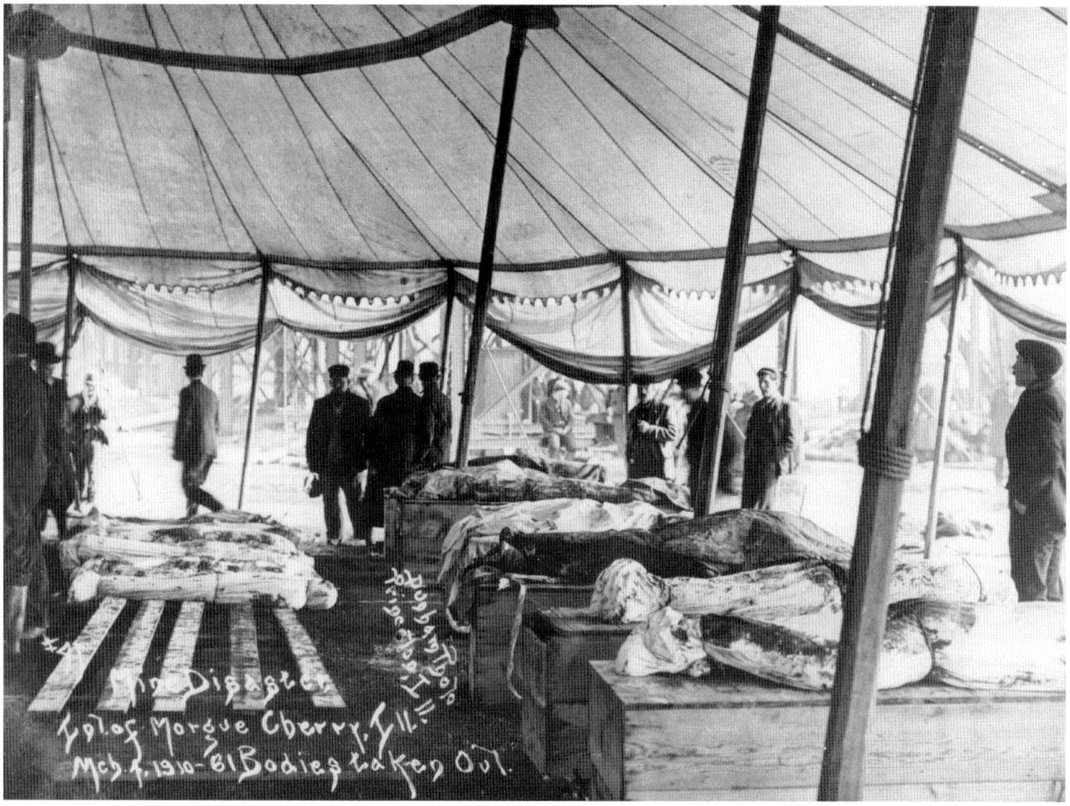

Bringing Up the Bodies and Identifying the Victims

Cherry Ill — Mar 4, 1910 — Churchill Peru

The Illustrated History of the Cherry Mine Disaster of 1909

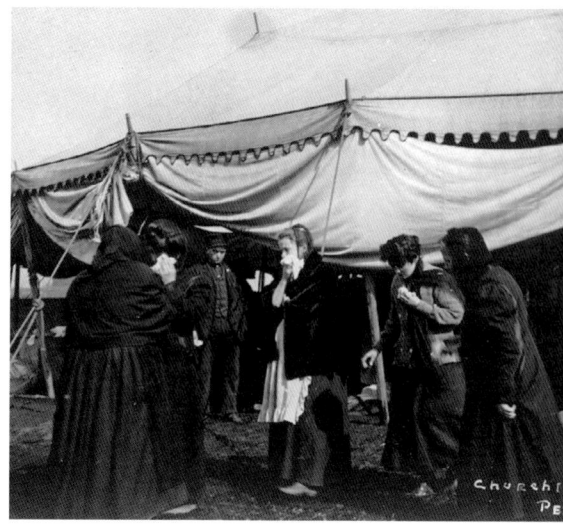

Photographers captured many moments of grieving women waiting for the news.

Bringing Up the Bodies and Identifying the Victims

Above left: Caskets lined up outside the makeshift morgue.

The Illustrated History of the Cherry Mine Disaster of 1909

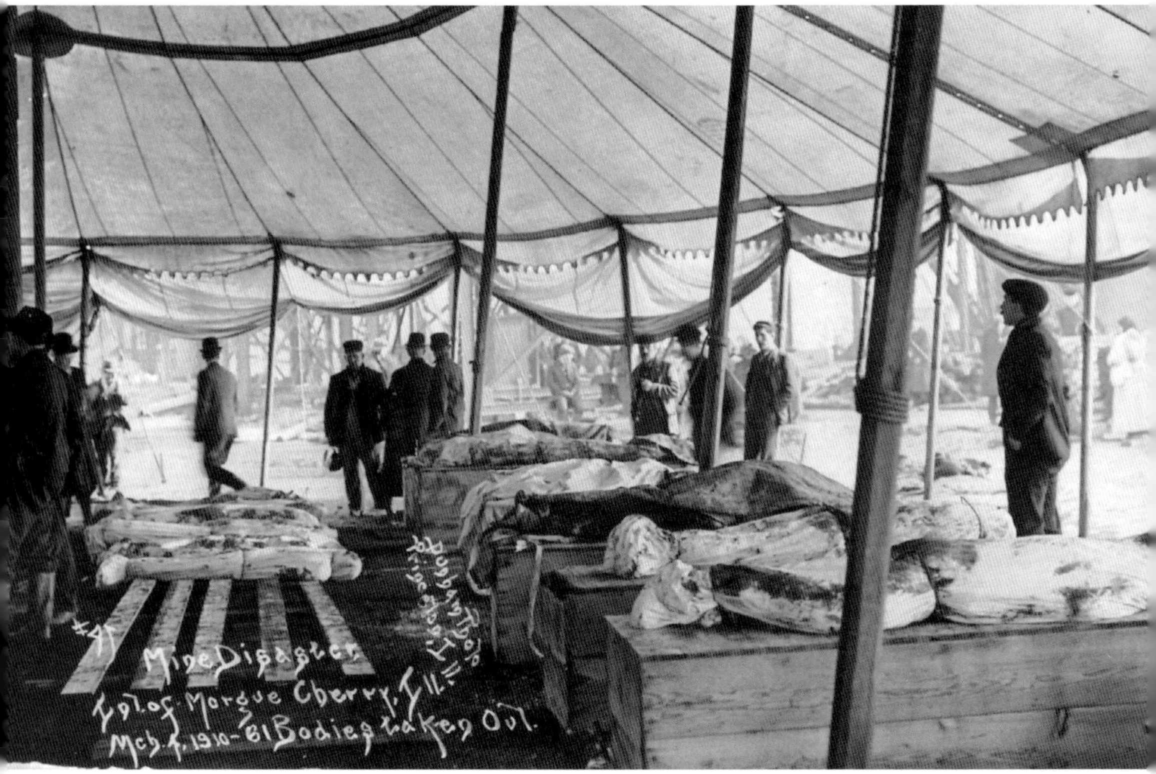

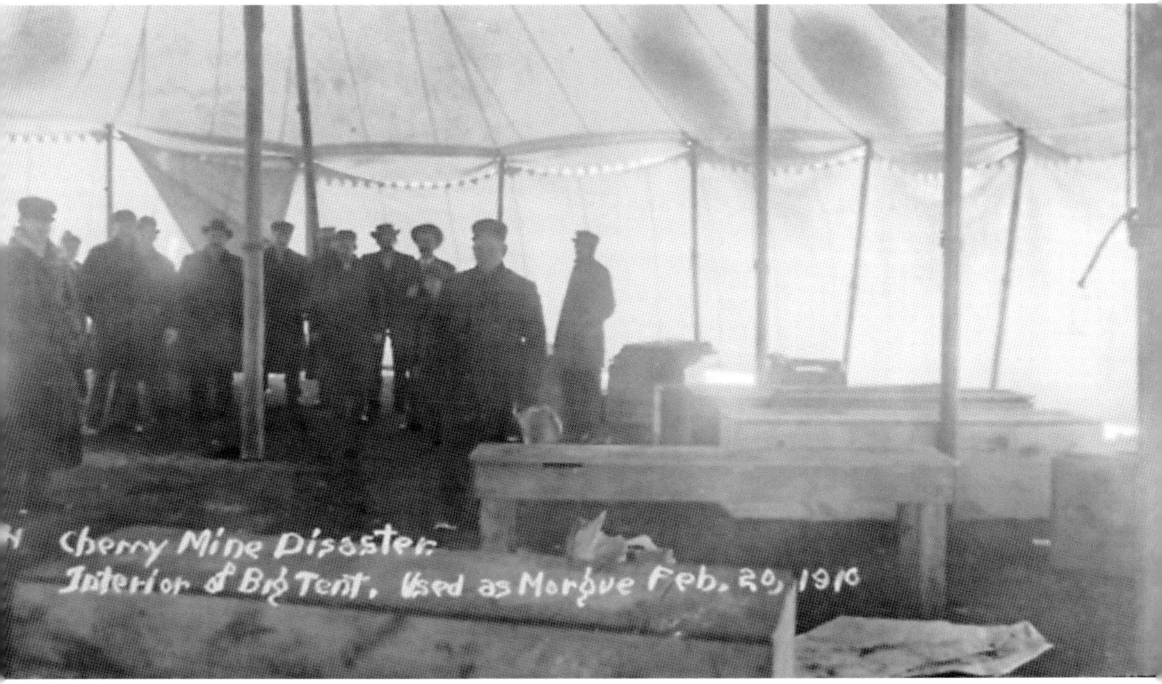

Bringing Up the Bodies and Identifying the Victims

The Illustrated History of the Cherry Mine Disaster of 1909

A tool shed, along with several tents, was used as a morgue when the bodies began coming up from the mine.

Cherry, Ill. Apr 12-1910
Churchill Photo
Peru Ill.

Cherry Ill. Churchill Peru.

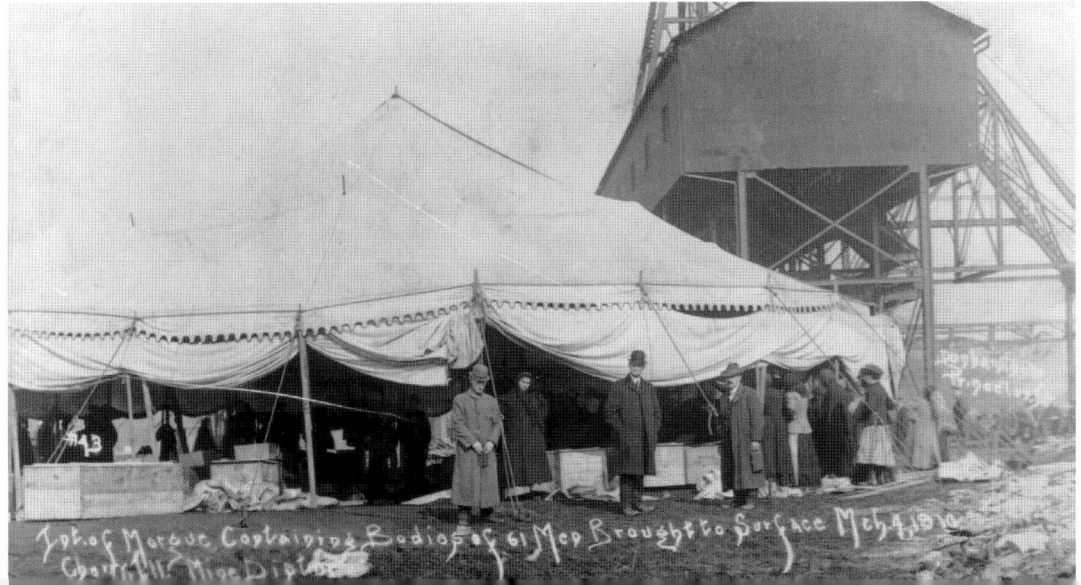

Ipk. of Morgue Containing Bodies of 61 Men Brought to Surface Mch 4, 1910
Cherry Ill. Mine Disaster

The Illustrated History of the Cherry Mine Disaster of 1909

122

Bringing Up the Bodies and Identifying the Victims

The staff who had charge of the mine victims.

Above left: A miner with a 2,000-lb chunk of coal.

Above right: Mine carpenters built stretchers for the rescue and recovery effort.

The Illustrated History of the Cherry Mine Disaster of 1909

The dead were carried on stretchers as the bodies were brought up from the mine.

5

THE FUNERALS

Funerals were held almost every day during the spring of 1910 as it took several months to bring up all the bodies and identify them. Some of the dead were buried in mass graves in the Cherry cemetery.

The dead included twenty-two people between the ages of fifteen and twenty; thirty-six people between the ages of twenty-one and twenty-five; fifty-five people between the ages of twenty-six and thirty; fifty-five people between the ages of thirty-one and thirty-five; thirty-five people between the ages of thirty-six and forty; sixteen people between the ages of forty-one and forty-five; twelve people between the ages of forty-six and fifty; fourteen people between the ages of fifty-one and sixty-two; and eleven people with no ages reported.

The state report in 1910 gave the total for the dead: seventy-three Italians, thirty-six Slavs, twenty-eight Austrians, twenty-one Lithuanians, twenty-one Scots, fifteen Germans, eleven Americans, twelve French, nine Swedes, eight Poles, seven Belgians, three Irish, three Russians, two Greeks, and two Welsh. There were sixteen nationalities, 161 married men, and ninety-seven single men.

There were 160 widows and 470 orphans (407 under the age of fourteen) left after the tragedy. Thirty-three children were born after the deaths of their fathers. Twenty-two widows remarried and thirty-five widows went back to Europe.

There was a father and three sons killed: Joseph Timko, Sr., age fifty-one; Joseph Timko, Jr., age twenty-eight; Stephen Timko, age twenty-four; and Andrew Timko, age seventeen. Four Love brothers died: John, age thirty-four; Morrison, age thirty-one; David, age twenty-four; and James, age twenty-six.

Henry Kroll, age fifty-six, and his sons Alex, age twenty-three, and Alfred, age fifteen, died in the mine. James Leadache, age forty, and his sons, Frank, age twenty, and Joseph, age fourteen, also died. Three members of the Klemiar family died: George, age fifty-six; Thomas, age fifty-five; and Richard, age twenty-four.

There were other families with more than one person killed in the disaster: Josif Malnar, age thirty-nine, and his son, Slavoljub "Lewis" Malnar, age eighteen; Peter McCrudden, age forty-eight, and his son, John McCrudden, age twenty-five; John Franciskovic, age forty-eight, and his son, August Franciskovic, age twenty-three; Archie

The Illustrated History of the Cherry Mine Disaster of 1909

A funeral for three members of one family. This could be the Klemiar, Kroll, or Leadache family.

Marchiona, age fifty-two, and his son, Frank, age thirty-two; Michael Suhe, Sr., age forty-three, and his son, John Suhe, age seventeen; Alexander Norberg, age thirty-seven, and his brother, August Norberg, age thirty-four; Sam Howard, age twenty, and his brother, Alfred, age fourteen; George and Peter Atalakis; Paul and Charles Amellani; Peter and Antonio Bolla; Frank and Lewis Bawman; Charles and Joseph Budzom; Thomas and John Brown; Lewis and Angelo Costi; Canical and Angelo Ciocci; Charles and Eric Erickkson; Frank and Joseph Kornovia; Selcomo and John Lonzotti; Martin and Alfred Ossek; Perys and Peter Prusitus; John and Alex Pearson; Martin and Joseph Repsel; Edward and Paul Seitz; and Edward and Arthur Mills.

There were more miners who were not brothers or sons but were cousins or otherwise related to other miners.

The four Love brothers were related to Robert and Alexander Deans. Their sister, Jessie, was married to John Love. The Loves and Deans had breakfast together at John and Jessie's house before heading to the mine on November 13. Jessie later married John Fraser, an assistant mine manager at Cherry who had been married to Jessie's sister, Janet, in Scotland.

Harry Stewart was one of the twelve heroes who burned to death while rescuing fellow miners. His widow, Janet, gave birth to their fourth child, Robert, five days after Harry's death. She later married William Love, the only one of five brothers who survived the mine fire. Janet had four more children with William. To add to Janet's tragedy, her son, Robert Stewart, was killed in a mine accident in Eagerville in 1939.

On November 15, 1910, one year and two days after the fire started, a funeral was held for John Suhe, the last victim to be buried.

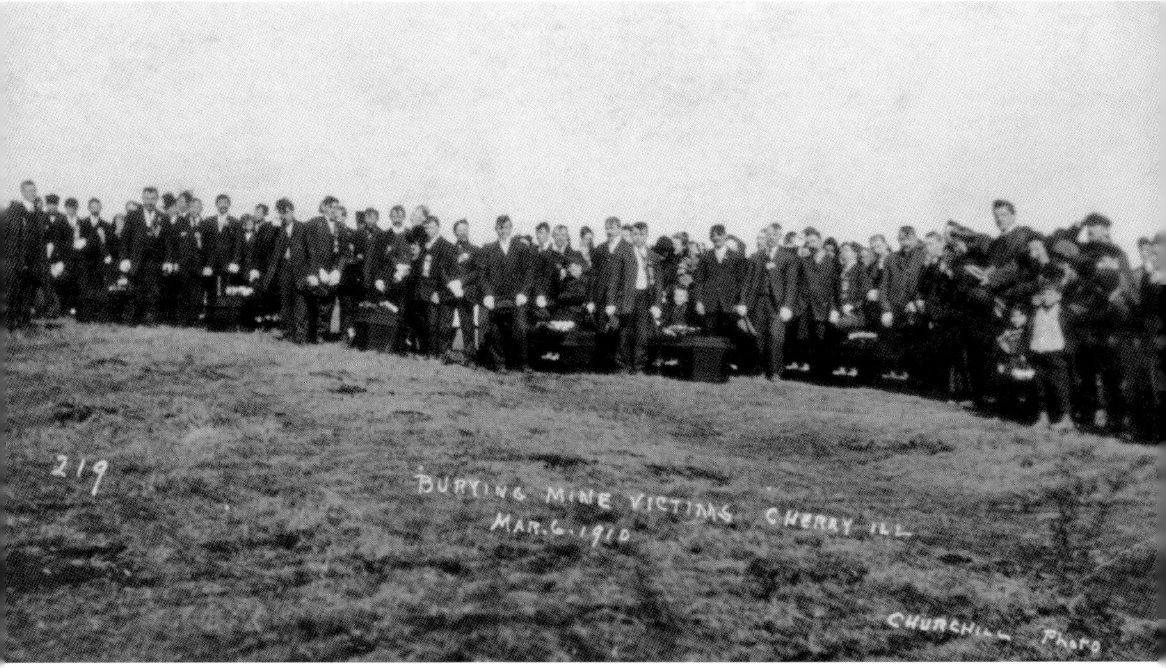
219
BURYING MINE VICTIMS CHERRY ILL
MAR.6.1910
CHURCHILL Photo

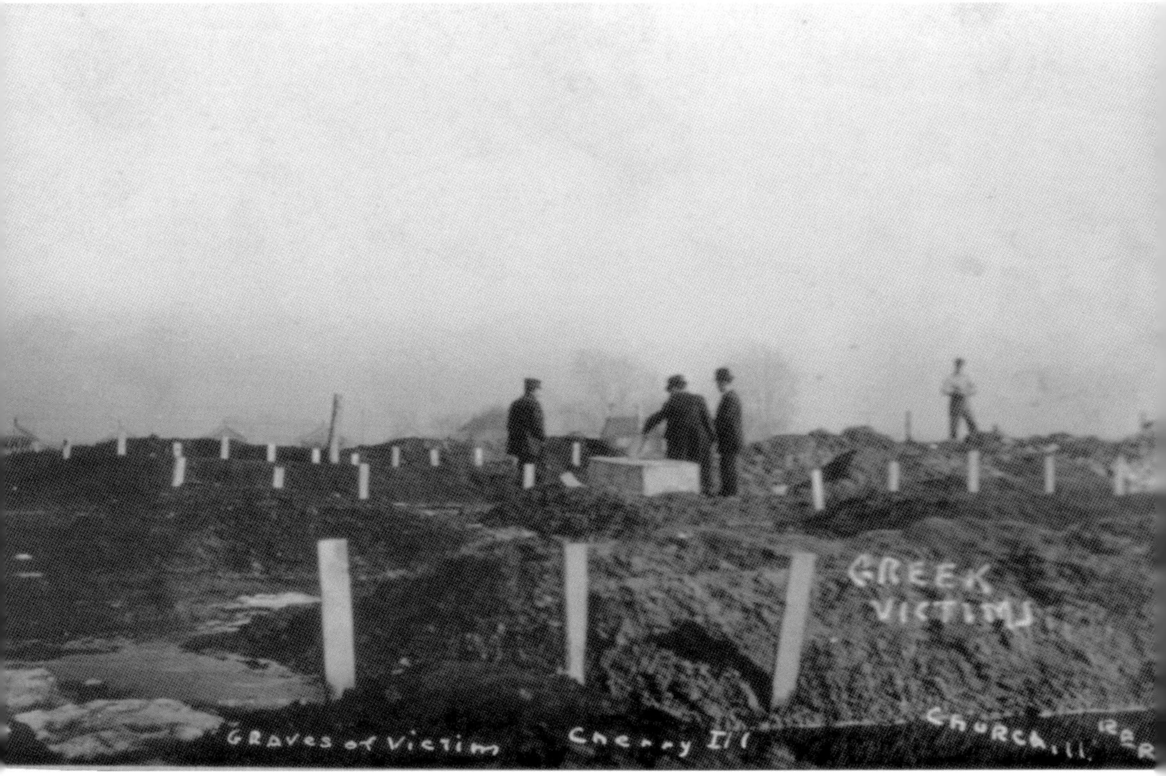
GREEK VICTIMS
GRAVES of Victim Cherry Ill
Churchill

The Illustrated History of the Cherry Mine Disaster of 1909

Many funerals were held at the Congregational Church and at Holy Trinity Catholic Church, on Cherry's main street. The Catholic church building was razed and rebuilt. The protestant church building today is United Church of Christ.

The Funerals

Funeral of L. Smith, a Cherry Mine Victim

The funeral of John Smith.

The Illustrated History of the Cherry Mine Disaster of 1909

The casket of John Davies is being readied to be shipped for burial at Kingston Mines, Illinois. The sixteen-year-old boy was hired only two days before being killed in the fire.

Viewing the remains of John Davies.

The Funerals

The Italians held elaborate funeral processions for the large number of funerals done at the same time.

The Illustrated History of the Cherry Mine Disaster of 1909

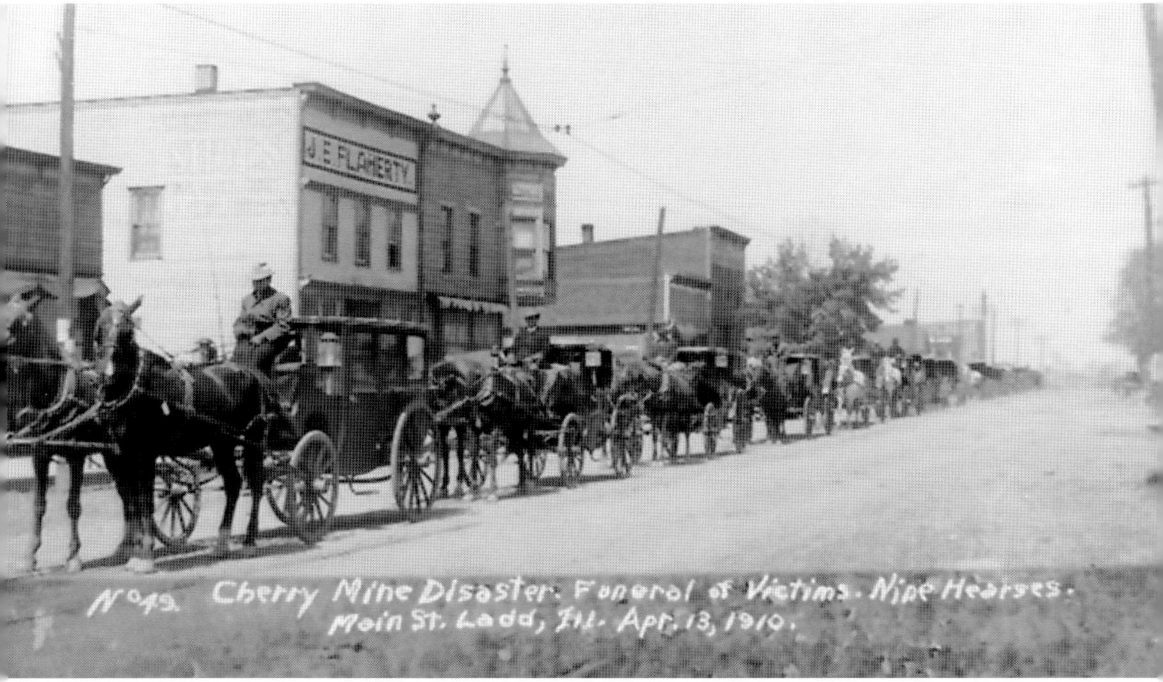

The Funerals

N°49. Cherry Mine Disaster: Funeral of Victims. Nine Hearses. Main St. Ladd, Ill. Apr. 13, 1910.

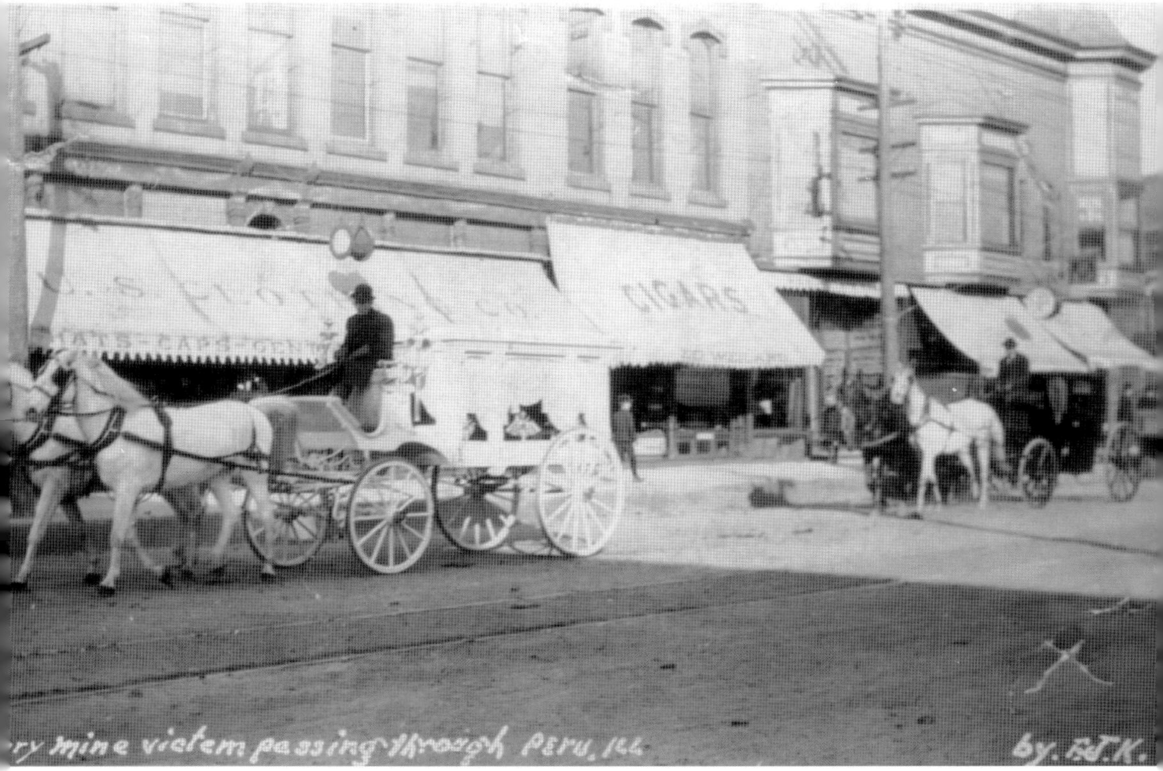

ry mine victim passing through Peru, Ill.   *by Pat K.*

The Illustrated History of the Cherry Mine Disaster of 1909

Funeral of two cherry mine victems, by F.S.K. Peru, Ill. No 1

The Funerals

Funeral of Father & son, two Cherry Mine Victem's PERU ILL

The Illustrated History of the Cherry Mine Disaster of 1909

Leopold Dumont.

The funeral for Leopold Dumont.

The Funerals

The Illustrated History of the Cherry Mine Disaster of 1909

The Funerals

KNIGHTS OF PYTHIAS AT CHERRY APR. 12-1910
ATTENDING FUNERALS OF 5 BROS. OF THE ORDER
241   IN THE SHADOW OF THE FAN
Churchill Photo.

Members of the Knights of Pythias lodge at the funeral of five lodge members on April 12, 1910. A double-exposure or a window reflection shows the shadow of the fan used in the mine by the "Eight-Day Men."

K. of P.s at Cherry Mine

Funeral for another Knights of Pythias lodge member.

Wreaths "At Rest" from the Knights of Pythias lodge.

The Funerals

A sad procession follows a mine victim to his final resting place.

The Illustrated History of the Cherry Mine Disaster of 1909

Two photographs showing funerals for seven people at a time.

The Funerals

CHERRY ILL
MINE VICTIMS
BURRIED AT PERU ILL
NOV. 25-09
CHURCHILL PHOTO
PERU IL

CHERRY ILL APR-12-1910
Churchill Photo

The Illustrated History of the Cherry Mine Disaster of 1909

Amid all the funerals for the miners, a commercial photographer added a photo for the funeral of a baby from the Henry and Margaret McClusky family.

22
CHERRY ILL
Churchill Peru Ill
Nov 14-09

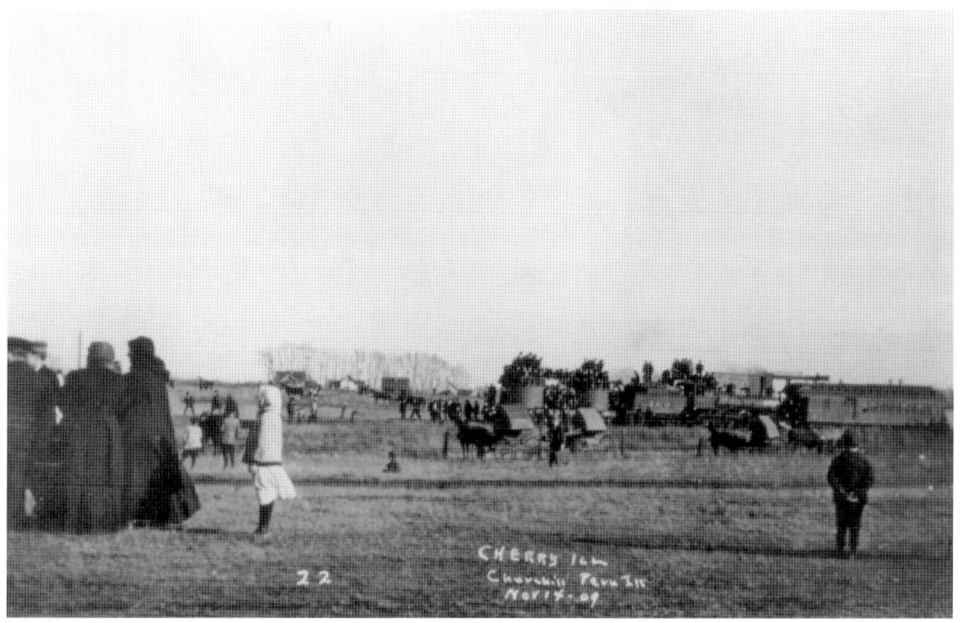

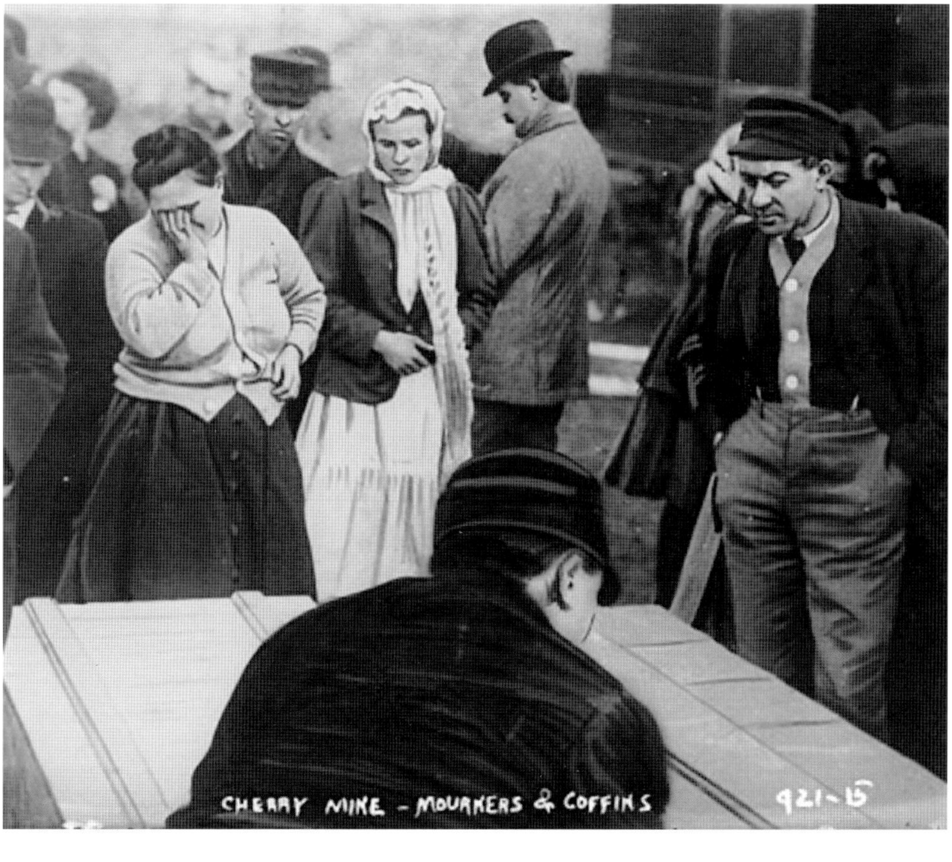
CHERRY MINE - MOURNERS & COFFINS
921-15

The Illustrated History of the Cherry Mine Disaster of 1909

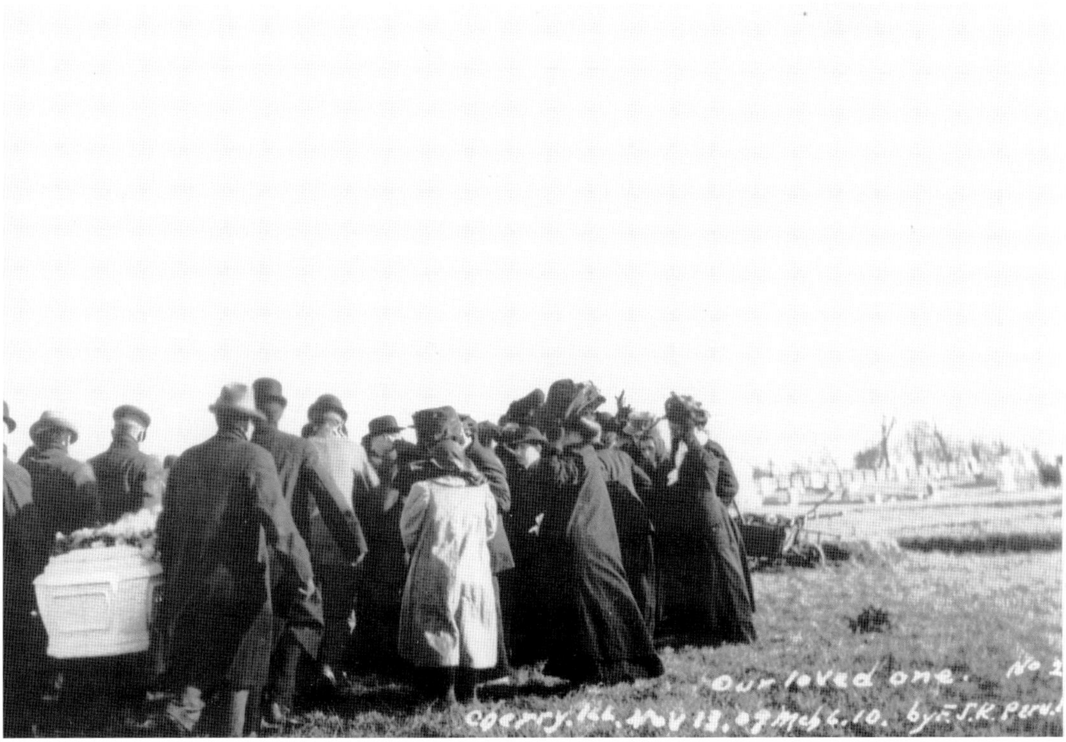
Our loved one.
No 2
Cherry, Ill. Nov 13, 09 Mch 6, 10. by F.S.K. Peru,

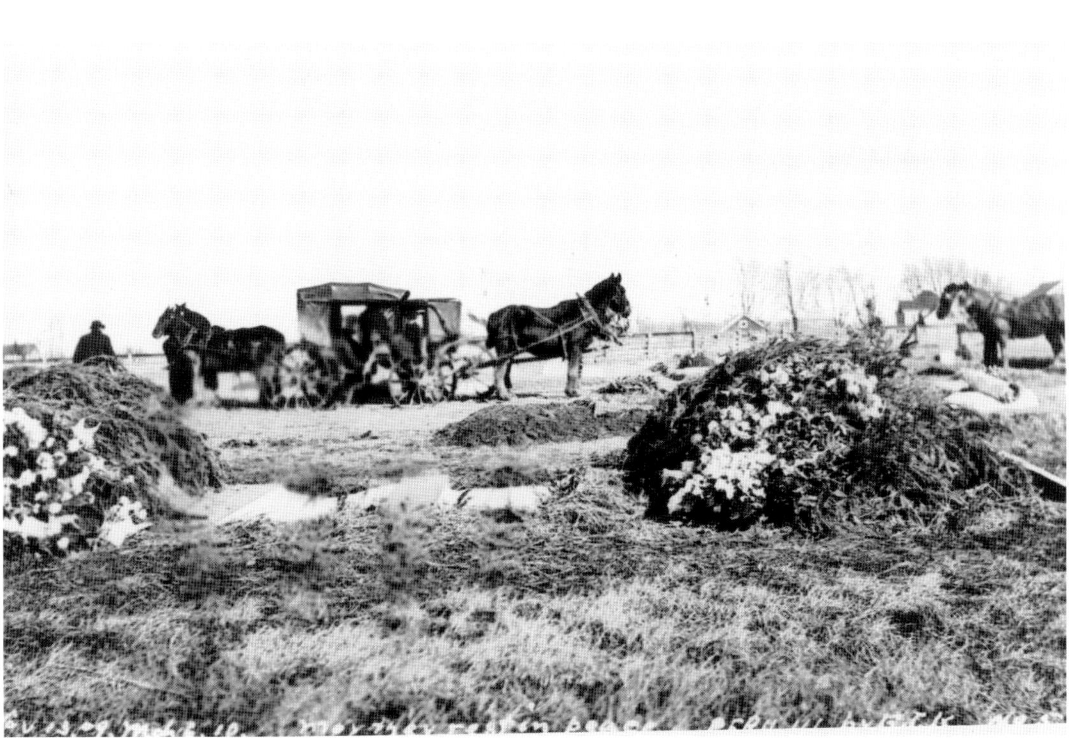

The Illustrated History of the Cherry Mine Disaster of 1909

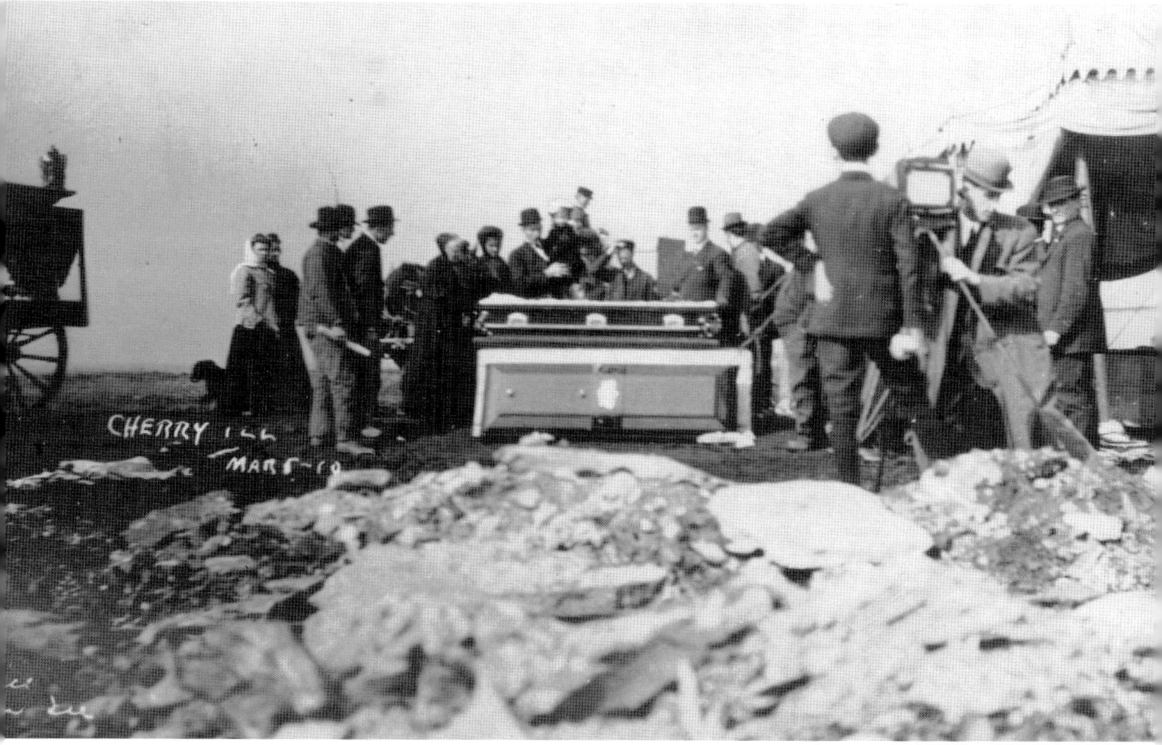
CHERRY ILL
MAR 5-10

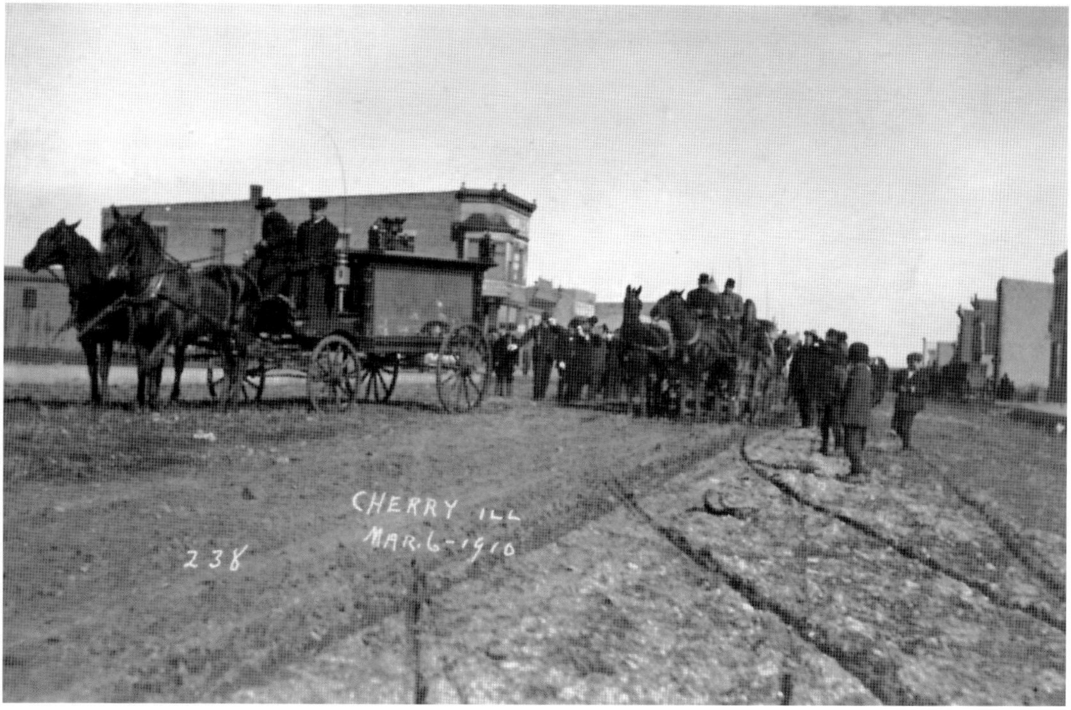
CHERRY ILL
MAR.6-1910
238

The Illustrated History of the Cherry Mine Disaster of 1909

A funeral at St. Joseph's Catholic Church in Peru, Illinois, for a father and son.

The Funerals

Cherry Mine Disaster Funeral Feb. 20, 1910

Leaving the Morgue Cherry 3/5-10

The Illustrated History of the Cherry Mine Disaster of 1909

Above left: The grave in the Cherry cemetery for Joseph Timko, who was killed, along with his three sons.

Above right: Tombstone for Antenore Quartaroli, leader of the "Eight-Day Men." He died in 1918.

Below left: Thomas Bayliff, a former Cherry mayor who died in the mine tragedy.

Below right: Peter and Anton Bolla, Italian immigrants who died in the mine.

The Funerals

Above left: John and Domenica Compasso.

Above right: John Compasso's grave in Mt. Olivet Cemetery in Spring Valley, Illinois.

Below left: Sherman Kingsley, a researcher for *The Survey*, whose work effected the workmen's compensation laws that came out of this disaster, and Ernest Bicknell, national director of the Red Cross, on the scene in Cherry.

Below right: George Eddy in his lodge outfit.

The Illustrated History of the Cherry Mine Disaster of 1909

Elizabeth Psak, Mary Pavlick, and Mary Packo—all widows in the Cherry cemetery. Below are the worn tombstones of John Psak, George Pavlick, and Andrew Packo, 100 years later.

The Funerals

Charles Bernardini was buried in a cemetery in nearby Ladd. He left a widow and a two-week-old child.

A 1910 photograph of the marker for John Franciskovic (named as Francisco on the state list) and his son, August, in the Cherry cemetery in 1910.

The Illustrated History of the Cherry Mine Disaster of 1909

Grave site for Urbain Leynaud in the Cherry cemetery, in 1910 and in 2011.

Leynaud is in the foreground, with a cap and a pipe, at a gathering of fellow Frenchmen and Belgians at Cherry. His wife is behind him, holding their baby.

The Funerals

A grieving widow and family view the remains of their loved one.

Grieving family members visit the cemetery in Cherry in July 1910.

Digging trenches for the graves for a number of victims.

Shovels are seen in the dirt where trenches were dug for the caskets of many victims.

One of the first bodies brought up was Ole Freeburg, a thirty-five-year-old timber man, who had been scalded to death.

John Suhe's funeral on November 15, 1910, was the last funeral of the mine victims. The Congregational Church is in the background.

6

THE AFTERMATH

A coroner's jury convened on December 2, 1909. Sixty-two people testified at the inquest. Duncan McDonald, president of the United Mine Workers, and Hector McAllister, state mine inspector for the district, were there.

A mine carpenter told the jury he saw men running from the mine for two hours after the fire was reported, even though it usually took one hour for miners to clear the mine after a shift. He said all the men could have escaped if they had been notified when the fire started. A mechanic testified that the bell to signal an alarm did not work. He also said a rail car sat on a trap door in the second vein, blocking an escape route. A number of witnesses refused to accept any responsibility for warning the men or in fighting the fire when it still was early. It was believed that mine company officials were coaching witnesses in order to distance themselves from any responsibility.

Dr. L. D. Howe testified about the heroic efforts of rescuers and said he did not see anyone in authority who was in charge during the emergency.

Alexander Rosenjack, who had been cast as a villain, even though he had bravely aided in rescue efforts, was not at the inquest. He was the one who authorities wanted to question about how the fire started and about his actions to stop it at its beginning. Rosenjack had promised he would be there. The sheriff was looking for him when he failed to appear. Rosenjack was reported to be in a number of other towns, sometimes using a false name to escape detection. Robert Deans also left town and did not attend the inquest. It was presumed that the coal company paid his way so he could not testify at the inquest about how the fire started.

The inquest was adjourned, because the jury could not reach a verdict without the testimony of Rosenjack and Deans. The inquest was reconvened a few more times, with more witnesses testifying. A final verdict was reached on May 10, 1910. It concluded that Alex Norberg died of burns in a cage near the second vein while trying to rescue miners, and that his death was caused by a confusion of the signals regulating the movement of the cage.

The verdict on the death of the miners, using Thomas Bayliff as the example, was that he died by suffocation caused by the fire:

The Aftermath

Coroner's jury investigating the circumstances of the mine fire.

The Illustrated History of the Cherry Mine Disaster of 1909

The fire was caused by a pit car load of baled hay coming in contact with, or in close proximity to, an oil torch in the second vein of said mine ... and said fire was caused by the careless handling of said car load while in transit to the third vein of the shaft. And we further find that there was a great delay in notifying the men in said mine of the danger by reason of said fire in time to insure their escape.

That last sentence very well summed up the reason for so many deaths. The proceedings revealed violations of child labor laws, inadequate escape shafts, the use of kerosene instead of animal or vegetable oil, a violation of the law that required engineers at each engine house, explosives stored inside the mine, and other violations.

There were so many factors that could have avoided the tragedy or minimized the tragedy. The fire could have been extinguished if Rosenjack and Deans had acted quickly. More men could have escaped if they had been notified at once, instead of thirty minutes after the fire started. The miscommunication of the alarm bells doomed the twelve rescuers to horrible deaths. The giant fan was reversed to draw out the smoke, but instead it fanned the fire. The two mining shafts, which were closed to smother the fire, pushed the black damp to the trapped miners.

Adding insult to injury, *The LaSalle Daily Tribune* reported that on the pay day following the accident, women appeared at the pay window instead of their husbands, "the faltering hands of women and children trembled over the envelopes, which were damp with tears." One Scottish girl, not more than sixteen years old, got her father's paycheck. It represented 165,600 pounds of coal he had dug the previous fourteen days. His pay was $52.99. However, despite the tragedy, the coal company deducted money for the blacksmith work on his tools, his house rent, union dues, powder, and doctor's treatment. The girl went away with $33.47. She left the office and burst into tears.

The Aftermath

The coroner's jury enters the mine on April 19, 1910.

Packages for relief of the Cherry sufferers pile up in the train mail room.

The Illustrated History of the Cherry Mine Disaster of 1909

A relief store was set up in what appears to be a saloon.

A girl runs to the relief train that has just arrived.

The Aftermath

Cover of the *Christian Herald* magazine, December 1, 1909.

The Illustrated History of the Cherry Mine Disaster of 1909

Bureau County State's Attorney L. M. Eckert is on the right.

Waiting at the Cherry train depot.

The growing popularity of photography brought photographers to chronicle numerous disasters in the United States. Such theater presentations as this became common in the early twentieth century. At least two of the "Eight-Day Men," William Clelland and George Stimac, went on the lecture circuit. Clelland gave illustrated talks for three years. George Stimac was giving talks less than a week after his rescue. Stimac spoke at the Princess Theater in Peoria on November 26, and met a friend who had survived a coal mine accident at Hazelton, Pennsylvania, in 1904, where he and eighteen men were buried alive for five days. Stimac made appearances in other theaters across the state, dressed in the same clothes he wore when he was rescued.

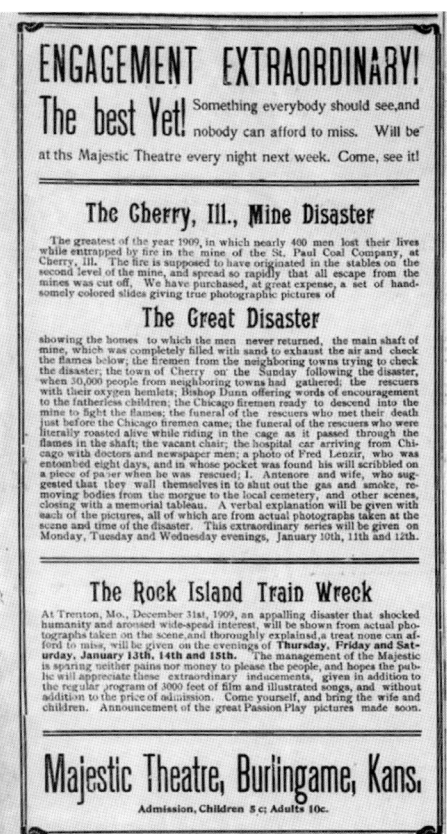

There are a number of other interesting items in the aftermath. Hundreds of men and dozens of mules below ground died, but rescuers found the rats in the mine survived the fire, the black damp, and the entombment. Miner Steve Timko was to go on trial for manslaughter for killing another miner, but his death in the mine, and the deaths of all fifteen witnesses, for and against him, ended that case. One mine inspector, Ted Fellows, ended up in the insane asylum in Kankakee and later committed suicide. Red Cross nurses reported that ninety of the widows were pregnant. Margaret Christopher of the Red Cross visited a home and found a mother and new-born baby, along with a priest there to baptize the boy; she was asked to be the godmother. Rev. E. J. Ridings visited the home of a widow with three children and an infant, who had come from Scotland just eight weeks earlier to join her husband.

The last survivor of the Cherry mine disaster died in 1990. Alexander Deans, who was a sixteen-year-old trapper at the time of the Cherry fire, died on February 24, 1990, in Cottonwood, Arizona, at the age of ninety-seven. He is buried in Gillespie City Cemetery, Gillespie, Illinois, next to his wife, Martha. Alexander was the brother of Robert Deans, who fled to Scotland rather than face questioning by the authorities.

Robert brought Alexander from Scotland two months before the Cherry disaster. They both sailed on the SS *Furnessia*, which left Glasgow, Scotland, on August 28, 1909. It arrived in New York on September 7, 1909. The ship, the largest in the world when it was built in 1880, was scrapped in 1911. The ship's manifest listed the men as brothers and as miners on their way to Cherry.

After the fire, Alexander Deans worked as a coal miner in downstate Gillespie until the late 1920s. He then worked as a night watchman and a maintenance man in Chicago before moving to Arizona sometime in the 1950s or 1960s.

Alex Rosenjack, the other mine worker who fled after the fire, eventually settled in Michigan. He died in 1957 and is buried in Memorial Park Cemetery in Battle Creek next to his wife, Lulu. Ironically, he spent his final years as a firefighter with the Battle Creek Fire Department.

The Cherry mine disaster had ramifications for the entire nation and for the coal industry. Federal regulations were enacted for what had been left to the states to regulate. Child labor laws were more strictly enforced. The Illinois legislature in March 1910 enacted stronger fire and safety regulations for mines. Engineers and other key workers had to be certified. More rescue stations were ordered. The state adopted a liability act in 1911, which later developed into the Illinois Workers' Compensation Act, the first such law in the nation. The United States Bureau of Mines was established in March 1910.

A second tragedy unfolded, with the compensation for the victims' families. Worker's compensation laws and employer liability were not yet the law. Lengthy negotiations went on until the coal company agreed to pay each widow $1,620. When funds from other sources were included —money from the Relief Fund, the United Mine Workers, Knights of Pythias and other sources—each family received a total of about $3,261.

For its penalty, the St. Paul Coal Company was fined $630 after admitting to nine violations of child labor laws.

7

THE PEOPLE

History is made not only of events and places, but also of the people who were there. This book gathers a little information from some of the families who lived in Cherry in 1909, both descendants of victims and survivors.

Henry Kroll, who died in the mine with his two sons, Alex and Alfred, was the great-grandfather of Wendy Winbeckler. She provided information for this book. She said that her grandfather, Ludwig "Lloyd" Winbeckler, an immigrant from Austria who came here to work in the Cherry mine, had a premonition on the morning of November 13, 1909. He told his wife, Marie (Kroll) Winbeckler, that he had a bad feeling about going to work. She prodded him, so he said he would go to work at lunch time. On the way there, he heard the alarm go off. He told Wendy the story years later, adding that his friends and family did not survive.

A story in *The Wide World Magazine* in October 1911, quoted Thomas White:

> In this attempt to reach safety, we lost Kroll and his son, and although we called repeatedly to them, so that they would know where we were and could find us, we never saw or heard anything from them. They were found dead in this passage to the escapement shaft after the mine was reopened, clasped in each other's arms. The old man had probably been overcome by the smoke, and the brave boy had preferred to die rather than leave his father to perish alone.

Dr. Lyston D. Howe, the mine company doctor, later moved to Streator and set up a medical practice. He died in 1942.

Francesco Zanarini was one of the men rescued after eight days. His brother, Riccardo Zanarini, avoided death by forgetting to bring his admittance badge to the mine that day. Francesco later opened a store in Cherry.

John Bundy, who was recruited by Mr. Cherry to be a mine manager in 1905, was one of the twelve heroes who died in rescue efforts. He began his career as a coal miner in Wales before coming to America and settling in Streator, Illinois. After his death, his widow, Sarah, moved back to Streator. She died in 1947.

John Navarro told a reporter from the *Kankakee Daily Republican* that he and his two sons, and his son-in-law, had entered the mine just after the fire got under way. The

The Illustrated History of the Cherry Mine Disaster of 1909

Alex Kroll.

Alfred Kroll.

Lloyd Winbeckler.

The People

Above: Lloyd Winbeckler married Marie Kroll in 1908. The wedding party included Alex Kroll, in the back row on the right, who died in the mine along with his brother and father. Pictured on the right are Marie and her son, Lloyd.

Right: Thomas Bayliff.

boys carried the older man over the bodies of dead and dying men and mules, through the dense smoke to one of the last cages up.

Dominic Formento, the local grocer who burned to death while rescuing other men, was born in Torino, Italy. His brother-in-law, Antonio Chiavario, was going to be one of the volunteers in the rescue, but Formento took his place, telling him to stay back and take care of the family. Dominic's widow, Julia, lived until 1960.

Thomas Bayliff was born on January 2, 1879. Bayliff was Cherry's first mayor, elected to a two-year term in 1905. His wife, Lillian, died on February 7, 1912. They had two children: Rose, who died in 1991, and John, who died in 2001.

Augusto Sergenti died in the third vein of the mine; he is misidentified on the state list as August Sarginto. His wife, Angiolina, gave birth to their third child, Augusta, twelve days after the disaster.

Two years later, on Augusta's second birthday, Angiolina married Luciano Galletti, a survivor of the Cherry disaster. After their son, Primo, was born, Luciano and Angiolina went back to Fanano, Italy.

Primo Galletti returned to America in 1929 and went to Cherry. He worked in the mine for half a day. "He worked till lunch time, and then quit, saying he did not think he would get out alive," his daughter, Lorena Cotton, said. Lorena and her husband, Dean, wrote a memoir of their family and the mine disaster: *Oneness Angiolina*.

At the age of seventeen, Primo moved to Ottawa, Illinois, and got a job at the glass factory at Naplate. He worked there for forty-three years.

Luciano's nephew, Luigi Galletti, married Beatrice Tintori and they moved to Braceville. He was killed in a mine accident there in 1905. Their son, John, was just fifteen when he was killed in the Cherry mine disaster. John Galletti's name is listed as John Garletti in the state report.

Giovanni "John" Tintori, born in 1888 in Fanano, Italy, survived the mine fire. He was a cousin of Giovanni Galletti. Giovanni arrived in America on October 23, 1909, and went to work in the Cherry mine. He survived the mine disaster.

John Tintori and his wife Catherine later moved to Detroit, where he worked in the automobile industry; he was the grandfather of Karen Tintori, the author of *Trapped*, the story of the Cherry mine disaster.

George Eddy was one of the true heroes of the disaster. He was a night boss, so he was not on the job when the fire started that afternoon, but he rushed to the scene and aided in the rescue efforts, getting trapped underground for eight days.

George Eddy was born in England in 1862 and immigrated to the United States in 1871. He married Elizabeth Bell in Streator, Illinois, in 1888. They had three daughters: Jennie (1889–1978), Florence (1892–1975), and Esther (1901–1994). Jennie married and had two daughters: Florence and Leona. Florence went to the University of Illinois and became an engineer; she never married. Esther married Merle Hatch in 1928 and they had one son, James Eddy Hatch, in 1936.

George Eddy died in 1919 at the age of fifty-seven. After the mine disaster, George did not work much because of a number of health issues stemming from his years as a miner and from conditions while trapped underground for eight days.

Edward and Arthur Mills came from England, and both were killed in the disaster. This was Edward's second journey in America. Several years earlier, he came to pan for gold in the Klondike, and he worked herding sheep. He went back to England, got

Angiolina Sergenti holds daughter Augusta behind the coffin of her husband. Three flowers on the box signify their three children. Their son, Armando, is next to her, held by an older boy. Beatrice (Tintori) Galletti, in the big hat, is next to the boys. On her other side is Mary (Galletti) Corsini, looking at Angiolina; Mary's husband, Attilio, is behind her. The little girl just behind Mary is Edith Amellani, whose father, Carlo, died in the mine.

Another view of Augusto Sergenti's casket, with his widow, Angiolina, standing behind it.

The Illustrated History of the Cherry Mine Disaster of 1909

The funeral of Johnny Galletti and Carlo Elfi. Angiolina (Galletti) Sergenti is holding infant daughter Augusta behind the coffin on the left. Luciano Galletti (her future husband) is behind her. Beatrice (Tintori) Galletti is the woman in the hat.

George Eddy and his family, in photographs provided for this book by his great-grandson, Alan Hatch.

married, and then returned to America in 1905 with his wife, Isabella, Arthur, and his wife, Alice. The bodies of Edward and Arthur were not recovered until April 1910.

Edward's daughter, Alma Buswink (1905–1996), told historian Jack Rooney that her father was the head of the local miners' union and they lived behind the school. She remembered that they had her father's wake in the house, and the men who carried the casket kicked the door a bit as they carried it in, and it scared her.

Mr. Rooney also spoke in the 1970s with Mike Phillips, the son of John Phillips, who escaped the fire. Mike's father said that he and his other son, John, were heading through the smoke on the east "run-around" toward the main shaft. They met Edward Mills, who was leading a group of men the opposite way toward the escape shaft. Mr. Rooney said:

> They had differences of opinion on the correct way out. I had the thought that if that were true, Mr. Mills' body would have to have been found near the escape shaft on the second vein. Sure enough, the coroner's log lists him as one of the first bodies found, right near the escape shaft. At that point in the 1970s, with the leaves off the trees, you could just about see Alma's house from Mike's back porch.

Frank James, one of the Cherry mine victims, came from Wales in 1906. His wife, Mary, and daughter, Daisy, soon followed. Daisy later married John William Brickhouse. Their son was legendary sports broadcaster Jack Brickhouse (1916–1998).

When the alarm sounded at the mine on November 13, alerting people to an emergency at the mine, Mary Jagodzinski was one of many people who ran toward the

Above left: Edward Mills.

Above right: Samuel Howard, who died just before rescuers could get to him.

mine. She was worried about her brother, Frank Jagodzinski (incorrectly named Frank Yagoginski on the state list of victims). Frank, age thirty-four, a Polish immigrant, was a driver in the mine. He was one of the first miners to come out, before the situation became hopeless.

He was frantic. In his limited English, he kept repeating, "Where's the kid? Where's the kid?" Everyone knew he was worried about his son, Frank, Jr., who was still in the mine. Frank, Jr., was only fourteen years old. Legally, he was not supposed to be working inside a coal mine, but the company broke the laws and bent the rules to employ children in the mine.

The others tried to restrain Frank, Sr., from re-entering the mine, but he broke away and went back down to find his son. Frank, Jr., came up, alive, in the next cage of men being rescued, but his father never came up. He died in the mine, burned to death by the fire. His body was found in a kneeling position. His wife believed he was praying when he died.

Frank's widow, Elizabeth (O'Donnell) Jagodzinski, and her five daughters moved in with her father. She later moved to Peoria and married John Hack, and they had a son, John, Jr. Frank Jagodzinski, Jr., married Helen Doerr and changed his last name to Jadd. He worked for the rest of his life as a projectionist at the Madison Theater in Peoria. He died in 1966. Mary Jagodzinski married Fred Tauscher. Their daughter, Clara, married Vernon Eustice. Vernon and Clara's daughter, Mary, married James Miller. Mary Miller provided the family story for this book. She continues to live in the Peoria area.

John Smith was a fifty-six-year-old English miner who died in the Cherry mine on November 13. Linda Sullivan said the family story is that John kissed his four-year-old daughter, Phyllis, goodbye that morning, which was something he did not ordinarily do before heading for work. Smith had cousins and other relatives in the mine. Some survived, others did not. Many bodies were difficult to identify after being pulled from the mine after so many months. John was identified by his shoes. When Phyllis grew up, she married James Armstrong, a nephew of Thomas White, who was one of the "Eight-Day Men" and later was mayor of Cherry.

James Steele was the mine superintendent at Cherry when the fire started. He was born in Scotland in 1866 and began working in the coal mines when he was thirteen. He came to America with his family in 1881. He worked in coal mines in Putnam and Bureau counties all his life.

He began working for the St. Paul Coal Company in 1903 as a mine surveyor and manager. The first coal shaft was sunk in Granville. Company houses for the miners were built, with the $7.50 monthly rent deducted from the miner's paycheck. By late 1904, the shaft was 500 feet deep into a third vein. St. Paul's mine at Mark, Illinois, was hoisting about 500 tons of coal per day.

Steele became superintendent of the Mark mine in 1905. He took over from James Cherry, who began sinking a mine for the company at a new spot that would be named for him. Operations in Cherry started on December 11, 1905, with 400 tons of coal hoisted on that day. James Cherry died on September 17, 1909. James Steele became superintendent of the St. Paul Coal Company. He held that job until the company closed in 1924.

It was James Steele who made the decision to seal the mine later in the day after the fire got out of control. He believed air currents coming down the air shaft was only fueling the fire, and shutting off the oxygen would snuff off the fire. It was a controversial decision, and many people thought he was condemning any miners below who were still alive. However, it probably stemmed the fire enough for rescuers to get into the mine eight days later to save twenty-one men who survived that long.

The Cherry mine was back in operation by late 1910. It had 494 men working, averaging 1,367 tons per day by 1913.

The mine at Mark achieved a state record in 1911 by hoisting a car load of coal every twenty seconds, or 2,247 tons for eight hours.

A fire broke out in the machine shop at the Mark mine on June 10, 1921. Jimmie Steele, Jr., discovered the fire. He was the head electrician and had left the office for a while when the shop caught on fire. He and Frank Holsinger tried to put out the fire, but Holsinger passed out from the heat and the gas, and Steele had to drag him out. Steele also had to be rescued. There were 500 miners underground. Since the cage could bring up only ten men at a time, it took a while to get everyone out. Spectators worried that this could become another disaster like Cherry. However, the fire was out by the next morning, with no lives lost.

The Mark mine closed in September 1924. The Cherry mine closed in April 1922 but soon reopened until the company closed it in 1927.

There was a John Brown and a Thomas Brown who both died in the mine fire. Another man, John Thomas Brown, survived the fire, as one of the "Eight-Day Men." He was born in England in 1885 and married Mary Ann Pierce in 1905. He came to America in January 1909 on the SS *Saxonia*, along with Frank Waite, who became his partner in the Cherry mine. Mary Ann and their children—Annie, Thomas, and James—arrived in Cherry in September 1909, just days before the mine disaster. The family moved to Des Moines, Iowa, in 1910. John Brown is sometimes misidentified as one of the victims. He died in Des Moines, Iowa, in 1967.

James Jamieson, a Scottish immigrant, was not in the mine when the fire started. He volunteered to go into the mine and rescue those who were trapped. He died there at the age of twenty. His body was one of the first ones brought up. He was identified by his father, Robert. He is buried in the Miller Cemetery in Spring Valley, Illinois.

The Illustrated History of the Cherry Mine Disaster of 1909

The rows of "company houses" for the miners were built by the coal company. The families paid the company rent and also had to buy goods from the company store. Most of what they earned in the mine went back to the company. The rows of houses here were named "Widow's Row" or "Dead Row" because almost every house saw the man of the house killed in the fire.

The People

Right: John Fasseo, who escaped when the fire started.

Below left: This young man is only identified as being one of the "Eight-Day Men."

Below right: James Jamieson.

Above left: Frank Waite and John Brown.

Above right: The children of vicitm Joseph Rodonis: Mabel, Joseph, Mary, and Alice.

Josip Malnar and his son, Slavojub (Lewis), were working in the Cherry mine when the fire broke out on November 13, 1909, and they became two of the victims that day. Josip Malnar was born in Croatia (Yugoslavia) and came to America in 1899 and found work in the coal mines near Rutland, Illinois. His wife Caroline and their children, Slavojub (born in 1891) and Mary (born in 1894), soon joined him. Between 1903 and 1908, they had four more children: Joseph (1903–1985), Anna (1906–1972), Eva (1907–1972), and Francis (1908–1995).

Josip's oldest daughter, Mary, had the sad task of having to identify the bodies for her distraught, pregnant mother. She identified her brother Lewis by his socks, which she had mended the night before the fire.

The family left Cherry for Rutland before the bodies were taken out of the mine. After the Malnar men were recovered, the family hired a wagon to take the bodies in coffins to Rutland. Unfortunately, the wagon broke down in Toluca. The driver said that was as far as he was going. Josif and Lewis were buried there, in a town where they had never been.

The People

The Josip Malnar family: Caroline, Joseph, Mary, Lewis, and Josip.

In July 1910, eight months after her husband died, Caroline gave birth to a daughter, who was called Elisabeth. Caroline died in 1963. Her last child, Elizabeth, died in 2004.

In October 1910, Mary Malnar married George Stimac. George was one of the twenty-one "Eight-Day Men" who were trapped in the mine when it was sealed for eight days. The family information is provided by Stephanie Schmidt, the great-granddaughter of George and Mary Stimac.

Juro "George" Stimac was born in Yugoslavia on April 24, 1887. He and his brother Frank came to the United States from Crni Lug, Croatia (Yugoslavia) around 1907. Even after experiencing the horror of surviving in a burning mine with no food and very limited water and air, George went back to coal mining until retirement, working in mines across central Illinois. George and Mary settled in Nokomis and had fourteen children. Mary (Malnar) Stimac died on April 20, 1958. George Stimac died in Springfield on April 10, 1963. He was the last survivor of the "Eight-Day Men."

The Illustrated History of the Cherry Mine Disaster of 1909

Above left: George Stimac.

Above right: George Stimac and Mary Malnar on their wedding day in 1910.

Main Street in Cherry after the disaster. Peter Monterastelli's store is the one with the awning. Delphina Dinelli, widow of miner Francisco Dinelli (incorrectly listed as Denalfi) moved to South Wilmington in 1910 and married Aeshal Monterastelli in 1911.

The People

Patrick Taggart, Sr., was the last man out of the Cherry mine, just behind his two sons (Patrick, Jr., and Ned) before the mine officials closed the airshaft in an attempt to cut off the oxygen that fed the fire. "When the smoke began to fill the mine, the miners were told to keep digging, as it was some one else's job to put out the fire," his great-grandson, Mike Kohr, said. "Grandpa Taggart would have none of that. He hustled his two boys up the airshaft. They lived; 259 men and boys died."

"Taggarts were very hard to kill," Mike continued. "It's hard to kill a man that thinks for himself. Moral of the story: Question authority. This was second nature to the Irish, who always had a very dim view of The Man in the Big House. Also: Defy stupidity."

Patrick Taggart was born in Ireland in 1861. He moved to Croy, Scotland, to find work and to escape the sectarian troubles in Ireland. The Protestant "Orange Men" murdered his father while he was "just mindin' his own business." In Scotland, he met and married Mary Mullholland. They moved to America. Mike Kohr said:

> Before arriving in Cherry, Pat Taggart rented a home next to a "sporting house" at the far north end of Greenwood Street, in the boisterous town of Spring Valley … One day after Patrick arrived home from work in the mines, Mrs. Taggart told her husband that she had been watching the house next door and could not figure out who the man of the house was. Mary, a devout Catholic, was having none of that. They moved to Cherry, post haste.

Mary and Patrick had seven children. When a Mullholland brother-in-law and wife died from the Spanish Flu epidemic in 1918, the Taggarts took in their seven children and raised them all in a two-bedroom miner's shack on Steele Street in Cherry.

Mike Kohr said, "I remember asking Grandma Sarah where everyone slept. She gave me a sidewise look and said, 'Everywhere.'"

Patrick Taggart, Sr., was Cherry's oldest resident when died on St. Patrick's Day 1942. Mary died in 1939. Patrick, Jr., died in 1938.

Patrick and Mary Taggart.

The Illustrated History of the Cherry Mine Disaster of 1909

This is the Taggart-Mulholland clan arriving at a site just northwest of Cherry in 1906. Sarah is in her father's lap in the center buggy pulled by the white horse. She was one year old and had been conceived on the voyage over from Ireland. "She said it was more than waves that were rocking that boat," grandson Mike Kohr said.

Above left: Sarah Taggart and her first husband, John Kohr, with their first-born child, Katherine. John died in 1936. Sarah died in 1998, six days before her ninety-fourth birthday.

Above right: Sarah (Taggart) Kohr, daughter of Patrick and Mary, is pictured boiling clothes outside the miner's shack on Steele Street that her father bought her for $78 after her husband, John Kohr, died in 1938. She raised six children in that house. The house stayed in the family for decades. Sarah moved two houses north, to her parents' home, after they died.

The People

John Flood was one of the "Twelve Heroes" who died in the burning cage while rescuing others. His great-nephew, Tom O'Shea, wrote:

> John Flood was known as the Scotch Merchant ... He was in his store when he heard the whistles blowing, and he rushed to the mine. Thinking his brother, Jim, was trapped, he went down on the cage three times, bringing up miners. He burned to death. Jim was not in the mine at the time, and survived. There is no indication that his other brother, Thomas Flood my grandfather, was down there at the time. Tom Flood died in the Great Flu Epidemic in 1918, a few years before my birth.

Mike Kohr had these observations:

> Most of the miners' houses were moved from mine to mine. Many of these came from South Carolina, after a mine closed there. They were separated at the ridge line, put on a rail car and moved to the new mine site. On unaltered miner's shacks you will always see the front door is offset from the center line of the house. This was due to the load-bearing beams that supported the different sides of the house. The homes were owned by the mine company, as was almost everything else in town. Lucky miners were able to save enough money to buy the house.
>
> In one "ghost town" here in Bureau County named Marquette, the mine closed up and all the miners were kicked out of the homes. The town sat vacant for a year before the mine company came back to move the homes and stores to another location. During that time, local people tore the buildings down and used the lumber to build new homes elsewhere. Local authorities looked the other way. The mine company was less than pleased.
>
> The mine sold the tools. If you needed the pick sharpened, you had to go to a company blacksmith to have it worked. Competition was not allowed.
>
> Miners were paid by the ton. The companies were notorious for rigging the scales. In Spring Valley, Illinois, the miners finally voted out the Republican bosses from City Hall. On their first day in office, they established an office of Weights and Measures. They went to the local mine with a certified scale and checked the first load up. The mine scale weighed it at 900 pounds. The certified scale from the new office of Weights and Measures pegged that same load at 1,700 pounds -- almost twice as heavy as the mine scale registered. The mining company blamed excessive dust on the workings of their scale for the error of 800 pounds in their favor.
>
> Cherry is a town of quiet everyday heroes, from the men who died in the cage to the women and men that work to keep their memory and the lesson of that day alive. My boyhood home, until I was five, was the house on the northwest corner at Park and Steele Streets, 'Widows Row.' I could not be more proud of the legacy of that old miner's shack.

Antenore Quartaroli was a leader of the "Eight-Day Men" who miraculously survived being buried alive in the burning mine. He was born in 1883 in Boretto, Parma, Italy. He and his brother, Paradiso, came to America and went to Iowa. Paradiso remained there, and Antenore went to Cherry.

Antenore married Erminia Castelli when she was seventeen. They had four children: Anton in May 1909, Geno in May 1910, Alberto who died twenty-eight days after birth in 1912, and Olga in August 1913.

The Illustrated History of the Cherry Mine Disaster of 1909

John Kmetz (listed as Kometz on the state list) and John Forgach lost their lives in the mine disaster. Mrs. Kmetz was the mother of Mrs. Forgach. Mrs. Kmetz went to Cherry to help her daughter pack her belongings. Mrs. Forgach and her two children went to live in the Kmetz home on East Main Street in Streator.

Antenore Quartaroli.

The People

Erminia was pregnant with her second child at the time Antenore went through his ordeal trapped underground in the mine. Despite his traumatic experience, Antenore went back to work in the Cherry mine after the disaster. Antenore, Erminia, and their two sons went to Italy in 1913 to sell some property. It was there that Olga was born, in Bologna. They returned when Olga was three months old.

Antenore died on August 21, 1918. He went to Indianapolis for an operation for a ruptured appendix, and he died from an infection. He is buried in Ladd Cemetery. After Antenore's death, Erminia placed her sons, Anton and Geno, in an orphanage run by the Knights of Pythias for the children of its members. Olga was sent to live with her aunt Pia Castelli's family in Cherry. The boys remained in the orphanage until Anton turned eighteen. Anton took Geno and Olga to Chicago after that. Olga went to beauty school and opened her own beauty shop.

Erminia went to Italy in 1918 to look for another husband. She married a man named Tonnini. They had a son, Oriano. She intended to return to Cherry to be with her children but instead remained in northern Italy with her husband and son. Erminia finally returned to America in 1945 after her second husband died. She went to work at the Salerno Cookie Company in Chicago and married Mike Vena, her third husband.

Olga married Frank LaRocca in 1936. They had two children: Carmelita in 1938 and Robert in 1943. Erminia died in 1970, within six weeks of Frank's death. Olga died in 1984. Antenore Quartaroli's granddaughter, Carmelita Collins, provided information for this book. Today, in 2020, she and her husband Peter live in rural Wilmington, Illinois.

Erminia Quartaroli.

Erminia and daughter, Olga.

The Illustrated History of the Cherry Mine Disaster of 1909

It was common in the late nineteenth and early twentieth centuries to take one final photograph of a loved one in their casket. This is Antenore Quartaroli in 1918.

Above left: Erminia and Olga.

Above right: Olga Quartaroli married Frank LaRocca in 1936.

The People

Above left: Erminia and her second husband, Mr. Tonnini, and their son, Oriano, in Italy.

Above right: Frank and Olga LaRocca in 1964.

Above left: At the Cherry Memorial dedication in 2009: author Karen Tintori, Richard Quartaroli (Antenore's grandson and Anton's son), Carmelita Collins (granddaughter of Antenore and Erminia Quartaroli and daughter of Olga) and Geraldine Shannon (granddaughter of Antenore and daughter of Anton).

Above right: Carmelita Collins in 2020.

The Illustrated History of the Cherry Mine Disaster of 1909

Cherry widows and orphans.

Cherry Public School, seen here a few years after the mine disaster.

The People

Children left fatherless by the mine disaster of November 13, 1909.

The Illustrated History of the Cherry Mine Disaster of 1909

The People

William Clelland, another of the "Eight-Day Men," went on the lecture circuit for three years after the fire, giving illustrated talks about his experience. He continued mining coal until 1925, when his back was broken by slate falling on him. William died in Peoria in 1940; his wife, Ellen, died in 1979. Their daughter, Frances, was born in Clarke City (Kankakee County) in 1901 when her father worked in the mine there. Francis was twenty-seven when she began working as a servant in the home of the forty-year-old widowed George Harris in Fairbury. They married in 1934. George died in 1965. Frances died in 1990.

Giovanni "John" Compasso worked in the coal mine in Dalzell, Illinois, until the middle of 1909, before moving to Cherry and becoming one of the Cherry victims. His body was among those recovered between April 10 and 12, 1910. John's body was identified by his brother-in-law, Thomas Buffo, by his tobacco pouch and watch.

Compasso was born in 1876 in Pratiglione, in the province of Torino, Italy. John married Dominica Buffo, the sister of Dominic Buffo, who helped build Torino, Illinois, in Will County, where he owned a saloon, a dance hall, a grocery store, and a boarding house. Torino was another coal town that was prosperous for many years until the mine closed. Today, the site of the former town is covered by the cooling lake of a Commonwealth Edison-Exelon nuclear power plant near Braidwood.

The Buffo family came from Pratiglione, Torino, Italy, in the early 1880s to work in the coal mine at Diamond, Illinois, in Grundy County. Diamond was the scene of another coal mine disaster, on February 16, 1883, when a flood took sixty-nine lives. John and Dominica Compasso had five children: Anna, Frank, Minnie, Louis, and a stillborn child.

In 1981, at the age of seventy-eight, Anna Compasso gave an interview to the *Joliet Herald-News* about the mine tragedy. "It was a Saturday and a beautiful Indian Summer day when my world and that of many others was shattered." The family agonized every day until her father's body was found, five months after the accident. "We walked along

The William Clelland family.

Main Street and there were funerals every day. The morbid music is what lingered in my mind, and still does.... I guess it was at that time I decided not to marry and have to face what my mother did."

The Compasso family remained in Cherry in a house the coal company provided for free to the families of the victims. The Compasso family moved to Spring Valley when Anna was ready for high school because Cherry had no high school. Anna said, "I attended high school for one month. I came home from school one afternoon and found my mother dead. She was a victim of the 1918 flu epidemic."

The family discussed who would take the children. "They sat listening, as the adults in the next room divided them, on the day of their mother's funeral," Peggy Lami said. Peggy, the granddaughter of Frank Compasso, grew up near Hennepin in Putnam County and now lives in Glen Ellyn. She provided the family information here.

Frank Compasso went to the family of his maternal uncle, Thomas Buffo, in rural Granville, Illinois. The two younger children, Minnie and Louis, went to the family of another maternal uncle, Dominic Buffo, in Torino, Illinois.

Anna went to live with a maternal aunt in Joliet, Rose (Buffo) Gaudio. Anna worked for Gerlach-Barklow calendar factory and later at Davidson's Cafeteria in Joliet before retiring in 1965. She died in 2002.

Louis later owned a farm between Reddick and Gardner, Illinois. He married Mary Naretta in 1934, and they had two children: Mary Lou and Michael. Louis died in 1995. Mary died in 2005.

John and Dominica Compasso and their first two children, Anna and Frank. On the right are the widow Domenica and her children: Ann and Frank in the back, Louis and Minnie in front.

The People

Minnie married Fred Cox in 1926. They had two children: Fred and Patricia. Minnie died in 1995; her third husband died in 1941. Minnie told the family story of her singing and dancing at her uncle Dominic Buffo's tavern in Torino (on the bar) to entertain friends and neighbors. Frank married Anna Nagra, the daughter of Thomas and Catherine (Bersano) Nagra, in 1926.

Peggy said:

His earliest memories were of delight for his father's singing (after a beer or two), and dislike of his mother's cooking, especially her bean stew … Indelible memories were set for Frank once his family moved to Cherry in 1909. Five-year-old Frank retained clear memories of horror, hoards of rescue workers and distraught survivors. Entire trains hauling the curious were followed by what seemed endless funerals and, finally, hunger and cold once Cherry businesses had neither food nor coal to sell.

School in Cherry was no fun for an outdoor boy like Frank … He often skipped school to go fishing, walking in the creeks and woods, or foraging for nuts, berries, or mushrooms. Many decades later, Frank realized his sister, Ann, entered high school while he, one year younger, was only in sixth grade. A relentless researcher, he visited the Cherry Grade School to discover that he had failed two grades—a memory not retained nor regretted.

Breaking a promise to his mother, Frank went down into the Mark (Illinois) coal mine to work for a short time until it was temporarily closed … He kept another promise to her by being a life-long teetotaler.

Above left: Anna Compasso.

Above right: Frank and Ann Compasso on their wedding day in 1926.

The Illustrated History of the Cherry Mine Disaster of 1909

Frank later moved to Joliet, where more of his mother's family lived.

With a promise of work, the boy who hated school enrolled in correspondence courses in algebra, geometry and trigonometry and passed the entrance exam for work as an apprentice at a steel wire mill. Soon he was a union carpenter, earning good wages for the first time.

Frank and Anna's only child, Catherine, was born in 1927 in Joliet. They lived there while rehabbing the Buffo house near Mark, Illinois. They moved into the house in 1939, and Dolly went to high school in Granville.

Catherine married Walter Urnikis. They had three children: Kathleen, Peggy, and John. Walter died in 1999, Catherine in 2009. Peggy said:

> He [Frank] said that every day that the mine was not sealed, his mother walked to the morgue tent to see if more bodies had been recovered. The three older children walked with her, but they did not go inside the morgue tent. Frank was old enough to go alone to stores early in the morning to collect wooden shipping boxes from rail deliveries to burn in their stove. He said that if he waited too late, all the wood would be gone.

He spent his later years researching the Cherry mine disaster and his Compasso relatives. He wrote letters to relatives and visited court houses, cemeteries, relatives, schools, and more to find his genealogy. He taught himself to speak, read, and write Italian, and he corresponded with scores of new-found relatives in America and Italy. He made the first of four visits to Italy in 1982, accompanied by his cousin, Mary Baldi.

"Standing atop a mountain near Pratiglione, Frank looked down on seven villages where his entire family began hundreds of years earlier. He reported that his feeling of being an orphan disappeared that day," Peggy said. Frank Compasso died in 1996. His wife, Ann, died in 1979.

Carlo Amellani (listed as Armelani in the state report) was working with his brothers, Paul and Pete, when the fire started. Pete said he smelled smoke and said they should get out. Carlo replied that someone would put out the fire. Pete was the only one of the brothers to make it out alive. Paul left a widow and five children. Carlo left a widow and four children; the youngest, Edith, was born just three weeks before the disaster. Carlo's widow, Antonette, was so overcome with grief that she never recovered. She died two months later, on January 13, of "a broken heart," a LaSalle newspaper story said. Carlo's body was recovered on March 4. Their son, Albert, was killed in a coal mining accident in 1921.

Camille Pierard was a miner. He stayed home the day of the tragedy to build a cistern at his house. He also was an amateur photographer. Camille and his twelve-year-old son, Jule, heard the whistle blow and they started toward the mine. Camille told his son to go back and get the camera. "That's how the very first picture was taken, and many more, as the fire progressed," said Jackie (Pierard) Uranich, Camille's granddaughter. She added:

> If you see numbers in the left bottom corner of the pictures, those are the ones my grandfather took. The others have the name of their studio on them. Other photographers developed their pictures in the Pierard pantry. The Pierards were a small part of the disaster story.

The People

This postcard scene went unidentified until now. It shows Ann, Frank, and Minnie Compasso outside the morgue tent on March 4, 1910, looking for their father. The woman is not identified.

Children waiting at the train depot for the milk supply to arrive. Food supplies were getting low.

The Illustrated History of the Cherry Mine Disaster of 1909

Another group of Cherry widows.

Above left: The children crying in this sad photo are identified as Carlo Amellani's sons: Albert, age four, and John, age two.

Above right: Camille Pierard.

The People

Above left: Antenore Quartaroli and his wife, Erminia.

Above right: Charles and Rachel Papet.

Above left: Leopold Dumont was killed in the disaster. The picture was taken by Camille Pierard in front of the Pierard home at 206 Second Street. The Congregational Church is in the background.

Above right: Rachel Papet, her children (Charles-John and Lucienne), and a neighbor girl in front of their Cherry house.

One of the miners killed, Leopold Dumont, boarded at the Pierard house. His wake was at the Pierard house, and he was buried in the miner's cemetery.

French immigrant Charles Papet died in the second vein of the mine. He was born in 1877 and came to America after meeting with Americans, who were in France buying mine equipment and machinery, and who were recruiting French miners with tales of good wages and plentiful work in the Cherry mine. Charles went to La Salle, Illinois, and later sent for his wife, Rachel, and infant daughter, Lucienne.

Charles' widow sent a letter from Cherry dated March 9, 1910, to her family in France (roughly translated):

> Dear beautiful brother and beautiful sister, I answer your letter to send you the sad news that I found my poor Charles. But you need to think the evil that it made me to see death. I've seen him dead, but it is impossible for me to believe and be consoled because he was not sick or hurt. I recognize his face. It did me good. There are men who are unrecognizable. I made the funeral March 4. I bought a place like this, I know it is for rest. In a few days, I will send you a map of the funeral of my poor Charles. I not tell you more. I'm a little sick. Will you embrace me well and the children.

After Charles' death, Rachel accepted passage from the St. Paul Coal Company and returned to France with her son and daughter in April. Rachel was in France when she received her husband's death certificate, dated May 6, 1910. She died one year later, in May 1911, leaving two orphans.

Charles' grandson, Roger Papet, of Nantes, France, has been in contact with Peggy Lami in America in recent years. She supplied a lot of information the family did not know. Roger wrote to Peggy:

> I do not know how to tell you all the happiness you gave me by sending me pictures of the memories. I have a sister and two brothers, older than me, that I sent my research and photos. They are really very surprised by this because, like me, they did not know how our grandfather was deceased in Cherry. Even our dad never knew the truth because we were always told that his father had died in a gas blast.

Four Love brothers died in the mine fire: David, James, John, and Morrison.

The family traces roots back to David Love, born in 1760 in Campsie or Lennoxtown, Scotland. He married Janet Morrison, and their son, Morrison Love (1780–1850), married Isabella Young (1779–1865). Morrison and Isabella had sons: John (1821–1887) and Morrison (1846–1904).

It was this Morrison who married Janet Kelly (1854–1940) in 1872 and had the sons who died in the Cherry mine. Morrison also was a miner in Scotland. Morrison and Janet (Kelly) Love had eleven children: Jane (1874–1922); John (1875–1909); Morrison (1878–1909); Janet (born in 1880); James (1883–1909); David (1885–1909); William (1888–1950); Thomas (1890–1891); Alexander (1892–1918); Agnes (born in 1895); and Archibald (1898–1963). Alexander died from tuberculosis at the age of twenty-five. Thomas died of pneumonia at eighteen months. Three of Archibald's daughters are still with us in 2020: Nessie, Jenny, and May.

The People

More orphans of the mine disaster—the girl holding the baby on the right is Anna Compasso, with her brother, Louis. The baby in front with the spotted coat, and the girl to the right are Charles-John and Lucienne Papet.

Widows and children of Cherry mine victims wait at the Cherry train depot. The woman in the rear holding the infant is widow Rachel Papet with Charles-John.

The Illustrated History of the Cherry Mine Disaster of 1909

Above left: The family of Morrison and Janet Love, including the sons who died in the mine fire.

Above right: John, Morrison, James, David, and William Love.

Above left: This picture in the Schlitz tavern on the southwest corner of Cherry Avenue and Main Street was taken on an anniversary within a few years of the disaster. The man on the far right is Frank Marchiando.

Above right: Janet Stewart and her children.

This is Joe Sandeen's saloon. Swedish immigrant Olaf Sandeen was killed in the mine fire. The short man with the hat on the left is Sam Howard, who also died in the mine. The tavern was a two-story brick building in the middle of the east side of the 100 block of North Main Street.

John Klaiser (listed as Klaeser on the state list) died in the mine. These are pictures of his family. The faces in the second picture reflect the terrible sadness following the death of the head of the family.

The Illustrated History of the Cherry Mine Disaster of 1909

Janet Stewart's husband, Harry, also was killed in the mine disaster. Janet gave birth to their fourth child five days after Harry's death. She later married William Love, who lived in Cherry at the time of the disaster. The family tree shows nearly a dozen men in the family with the first name of Morrison. Anne Love-Hoskins, the great-great niece of Morrison Love (1846–1904), provided the family information for this book.

John Klaiser's body was identified by his headlamp and check badge. There was a bone wrapped in a napkin in his pail, which he was bringing home for the family dog, as he often did. The Klaiser family was one of a handful of German Catholics in Cherry, and nearly all were affiliated with St. Joseph's Catholic Church in Peru. A number of those men were killed in the mine fire. The Klaiser children, Peter and Theresa, lived together in Cherry all their lives.

There were three company houses sitting by themselves northwest of the mine, and all belonged to German Catholics.

Edward Bruno, a thirty-three-year-old Italian immigrant, also died in the mine. His widow moved to South Wilmington in Grundy County, which was another coal mining town, just after the disaster. She went back to Italy in September 1910, taking her three children and the daughter of James Bruno of South Wilmington.

When such a monumental tragedy occurs, there are always people who look to place the blame. There was plenty of blame to go around for the Cherry tragedy, and a number of people were named. One was John Cowley, who was in charge of raising the cage when the Twelve Heroes perished while rescuing others. Even his obituary in the *Bureau County Republican* on March 16, 1916, noted he was operating the cage when the rescuers were burned alive. Renee (Piano) Toomey, the great-great-granddaughter of John Cowley, provided the story for this book.

As a very young child, I understood that something terrible had occurred in Cherry, and I knew our family had some connection to it, but I did not know any details. Occasionally, one of the elders in the family would try and discuss it with my grandmother, and she would definitely not want to talk about it, other than to say it was a terrible tragedy and it ruined a lot of lives. All throughout my childhood, on my grandmother's telephone stand was a large envelope marked 'Save' and underlined 'Grandpa,' stuffed with newspaper clippings. I remember being about ten or eleven, spending the summer with her, and picking up the envelope wanting to know more about what was inside. She told me there was nothing but sadness in there, and to put it down.

Later, when I was in high school in Naperville, Illinois, one of my classes discussed the Cherry Mine Disaster, and I was excited because I actually knew where Cherry was. That was the first time I learned about the details of the disaster. I initially presumed my grandmother's grandfather was one of the miners who perished in the disaster. Then, in 1984, at a 75th commemoration of the disaster, I read an article describing the hoist engineer as John Cowley. I recognized the family name, and the reluctance to discuss the event started to make sense. I asked my grandmother about it, and she said her grandfather had been treated very unfairly, and that he was just doing his job.

According to her, which she must have heard from her mother, John Cowley felt horrible for his role in the tragedy, and he never recovered. In their grief, many people blamed him for the deaths of the heroes, and he even had to leave the area for a period of time, in fear of his safety. The envelope my grandmother had contained all sorts of archival news clippings from the disaster, but I believe my grandfather threw the envelope away following my grandmother's death in 1992, because I did not find it among any of the household items when I cleared out the house following my grandfather's death in 1995.

I did, in the 1980s, have a conversation with my grandmother and her older sister Verna, on this subject. They told me that their grandfather was just following the rules, and that the company should have backed him up more than they did. My understanding is that he was eventually vindicated, but by that time, the damage had already been done, and people just wanted someone to blame. I was told I shouldn't tell anyone about my connection to John Cowley, because people might hold it against me. Of course, that warning seemed absurd to me. I lived and had a career in the suburbs, and doubted anyone in my orbit would care, but this was their mindset and probably explains my grandmother's reluctance and secrecy while I was growing up. So few other family members even know about this history, that I have verified the connection independently.

John and Mary (Edson) Cowley lived in Streator and had thirteen children. Eight were still living when John died, including a daughter, Mrs. Ralph Eddy of Streator, who was related to George Eddy. Five of their children lived in Cherry at that time.

Harry Parker was the first superintendent at the new coal mine at Cardiff, Illinois, in 1899. Cardiff is located southeast of Cherry, in Livingston County. He was there when the Cardiff mine blew up in 1903, which killed nine miners. He directed the sinking of a second mine, and everything prospered until the mine closed in 1912. The town of Cardiff vanished soon after that. Parker decided to go into a different line of work. In September 1908, Harry and his son, John, moved to another town to start a grocery business. The town they chose was Cherry.

Harry and John Parker were among the many citizens who assisted in the rescue efforts at the Cherry mine fire. Many of the Cherry miners traded at the Parker grocery store, and the Parkers let them buy goods on credit. The deaths of all those men who owed him money wiped out Harry Parker financially. He later moved back to Cardiff.

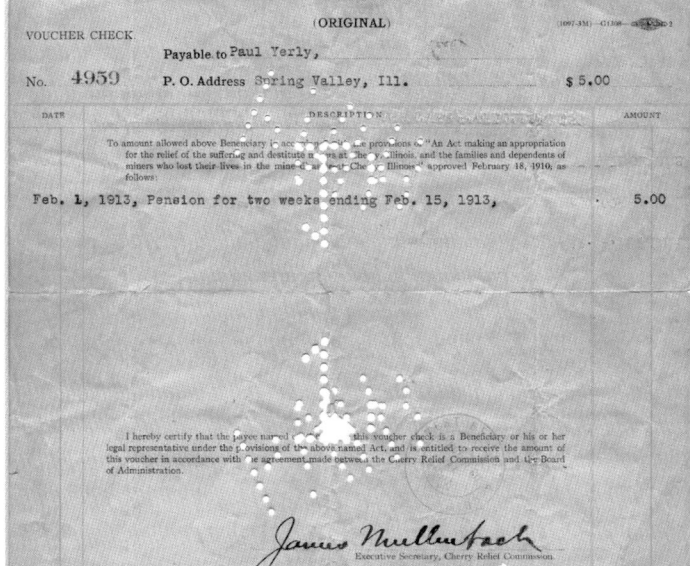

This is the voucher for a pension check made out to Paul Yerly for a two-week period in February 1913, for $5. It is signed by James Mullenback, executive secretary of the Cherry Relief Commission.

The Illustrated History of the Cherry Mine Disaster of 1909

Harry Parker's daughter, Eliza, married James Lettsome in Cherry in 1909. They had five children. The Lettsome men were miners, both in Cherry and Cardiff, and they were among those cited as helping in rescue efforts. James was nine years old when his father was killed in a mine at Braidwood.

Harry Parker's son-in-law, Thomas J. Williams, was a mine engineer at Cardiff. He later went to downstate Herrin and was killed on December 23, 1909. That day, an engineer taking measurements to extend the mine was told there was gas in the west entry. Williams, the assistant manager, said there was no gas. He led a group into the entry, where the flame from his lamp ignited the deadly gas. All eight men in that group were killed in the explosion.

There are a number of other connections between Cherry and Cardiff. Miestre Elario, one of the Cherry victims, is listed on the state report as living in Cardiff at the time of his death. Thomas Hewitt moved from Cardiff to Cherry, where descendants still live. John Flood, a Cherry businessman who died rescuing others, had family in Cardiff. Dick Cullen was a prominent citizen in Cardiff and later in Cherry. Several other families moved from Cardiff to Cherry to work in the mines or to work in businesses. Also, a number of Cardiff men went to Cherry to help with the rescue when the mine caught fire and later when it was reopened to recover bodies.

Louis Tintori worked in the Cardiff mine and later owned a saloon in Cardiff. John Ballotti married Mary Tintori and they had nine children in Cardiff; one of their sons was killed in a coal mine accident in downstate West Frankfort. There was a coal miner in Cardiff, Isadore Tintori, and his wife, Alvera.

Charles Atherton was a mine official in Cardiff until 1908, when he went to work at the St. Paul Coal Company mine at Granville. Atherton was transferred to the Cherry mine in 1909, helping in the recovery of bodies and repairs to the mine.

Atherton fired Milish Manditch, a twenty-three-year-old Austrian immigrant, on February 15, 1910. Three days later, Manditch went to the tipple, where Atherton was pushing a car loaded with props to a cage to be lowered into the mine. Manditch pulled a gun and shot Atherton in the shoulder. Atherton gasped, "What are you shooting at me for?" Manditch fired several more times, hitting Atherton in the abdomen. Dr. Howe was called from a nearby morgue tent and attended to Atherton before the wounded man was taken to a hospital in LaSalle.

Harry Parker is on the far left, with other company and state officials, following the explosions in the Cardiff mine in March 1903.

The People

Charles Atherton died on February 22. He and his wife, Bertha, had six children. Their seventh child, a son, was born on February 26, four days after his father died. Harry Parker showed up here again. He was an eyewitness to Atherton's murder and he gave a vivid account to the newspaper.

Half an hour before the shooting, Manditch went into F. C. Viner's store and bought a $6 revolver and ten bullets. He then went to Joe Sandeen's saloon, had a drink of whiskey, and left, saying, "Goodbye, Joe. In 20 minutes, I'm going to shoot somebody." After the shooting, Manditch was arrested in Joe Hozie's saloon. Hozie was a miner who barely escaped from the fire with his life. Manditch pleaded guilty to murder on April 16 and was sentenced to forty-five years in the state penitentiary at Joliet.

The tall tombstone for Alfio Amidei and Giuseppe Nanni (incorrectly named Alfio Amider and John Mani in the state list) in Ladd Cemetery had a long epitaph written in Italian, which roughly translates:

> To My Dear Husband Giuseppe Nanni, Died years 36, and My Son Beloved Alfio, Love of, Died Years 18, Another Dead in Catastrophe of Cherry on 13 Nov 1909. Archidamia and Daughter Maria Nanni and Sister Caterina Amidei Laid This in Remembrance With Great Affection.

It is not known if Giuseppe was related to Lorenzo Nani, a coal miner in Cardiff. Lorenzo's tombstone is somewhat similar, both in the design and the epitaph. As the Cardiff mine was winding down in 1911, Lorenzo went to work in a mine in downstate Rend City, intending to send for his family. However, he was killed in a cave-in there. His body was brought back for burial in Sacred Heart Cemetery near Cardiff. His tombstone, in Italian, reads: "Lorenzo Nani, died 17 July 1911, age 31. Cardiff, Ill. Left his wife with 5 children in bad circumstances."

Nani's wife, Orsolina, gave birth to another son, Lawrence Santini, four months after Lorenzo's death; he lived to celebrate his 100th birthday. He died in Joliet in 2013.

A tombstone in the Ladd Cemetery for three Cherry victims reads, "A Tear for Giulio Sorbelli and Gaetano and Francesco Ruggeri, Cruelly Lost in the Cherry Disaster on 13 Nov 1909. Leaving in Grief Parents, Brothers and Friends. Laid to Rest by the Widow Petronilla Sorbelli."

Charles and Bertha (Dachsteiner) Atherton and their children, Sadie, George, Marguerite, and Arnold.

The Illustrated History of the Cherry Mine Disaster of 1909

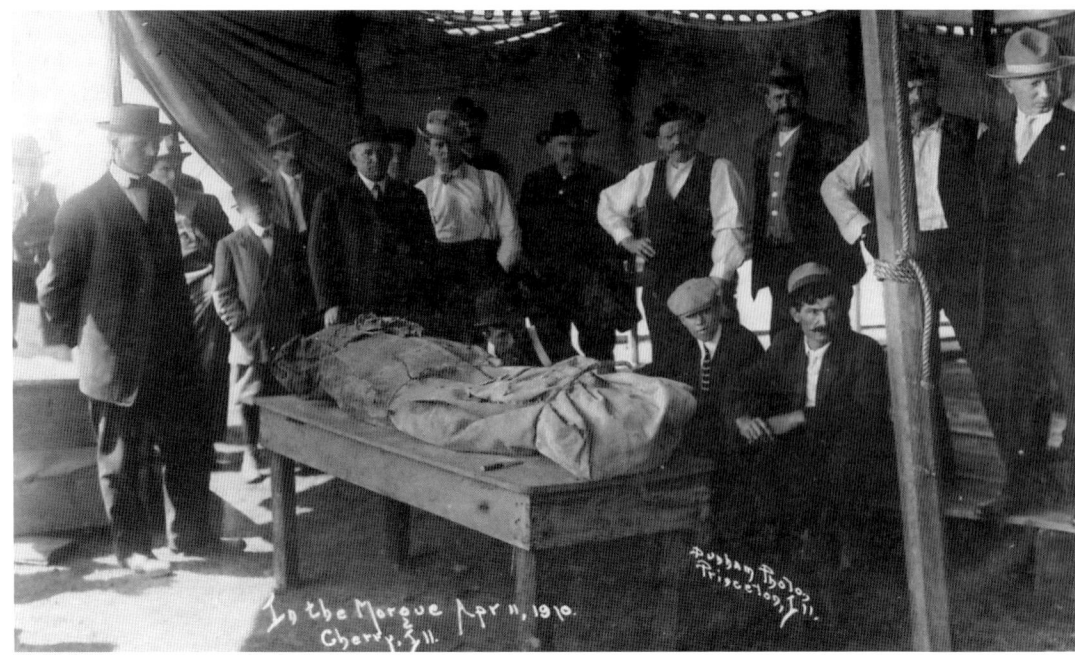

The Cherry–Cardiff connection provides some additional information to this story through personal postcards. We now have identification of two men pictured in a Cherry morgue tent, from the postcard Eliza (Parker) Lettsome in Cherry sent to her mother, Elizabeth Parker, in Cardiff. Harry Parker is the man in the suit just behind the body on the table. His son, John, is to the right in the white shirt and hat.

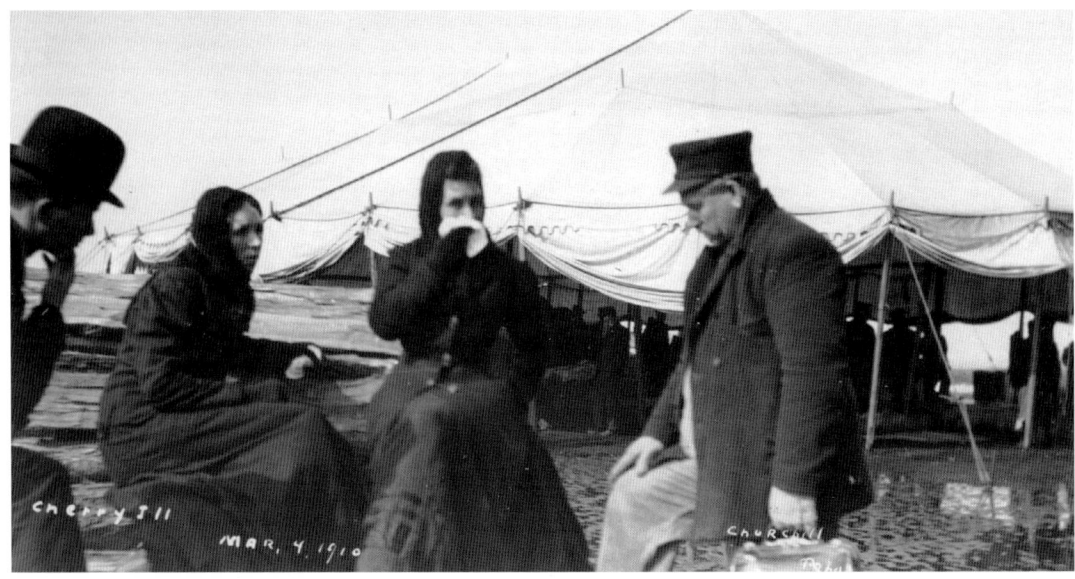

This postcard from Eliza to her mother in Cardiff has this message: "This is Mrs. White and Mrs. Miller, sisters from Iowa." George White and Edward Miller were among the victims.

This postcard sent by Eliza has this message: "This is them people that I took you to see. The middle one is the mother of the boy that was lost in the shaft. The old lady is her mother. Davis is their name." John Davies, sixteen, was one of the victims.

The message on this postcard from Eliza to her mother says: "Hello, Ma. How are you? That dark one is Tomas Bayliff. On each side of him is Bill Hyns and Paddy Richards, and other three—two brothers and father. Good bye. Write soon." Thomas Bayliff was thirty years old, and he left a widow and two children. William Hynds, a twenty-five-year-old miner, left a widow and one child. Thomas Richards was a twenty-one-year-old miner who left a widow. The two brothers and father could be Kroll or Leadache.

The Illustrated History of the Cherry Mine Disaster of 1909

This postcard sent by Eliza Lettsome in Cherry to her mother in Cardiff reads: "Look and see. Across Ike is Jim's brothers. The back one is Bayliff again, and the other is that Davis I was telling you about. Only worked one day. Only sixteen years old." Ike was Isaac Lewis, a liveryman in Cherry, who died while rescuing miners.

Unidentified men sit in front of a Cherry saloon with a statue of "Big Bill's Best Bitters."

8

REVISITING THE TRAGEDY

Recent examinations of the events have called into question some of the conventional theories about the cause of the tragedy and the motivations for the absence of key witnesses.

A fire in a mine is not uncommon and usually is not a cause for alarm if it quickly is put out. However, when the fire in the Cherry mine got past the point of being able to be extinguished by beating it with coats, almost every decision that was made was wrong.

Instead of bringing water to the burning car, the men tried to bring the burning car to the water. The tracks were blocked; that allowed the fire to spread to a wider area of the coal seam and the timbers. The fan was reversed, then shut off, causing more damage. It was one wrong decision after another.

It was accepted that Alexander Rosenjack and Bobby Deans fled Cherry to avoid taking the blame for letting the fire spread; to avoid retribution from a vengeful crowd; to avoid testifying and paying the penalty for their negligence, incompetence, and poor decisions; and to avoid giving testimony that would be damaging to the coal company. Cherry historian Jack Rooney believes they did not run but were spirited away by certain people in charge who did not want the truth to come out.

An important part of a miner's life was membership in the Knights of Pythias lodge. In those early days, with not all mines unionized, lodge membership was important. Lodge members got preferential treatment in the industry. Mr. Rooney has the lodge charter for Cherry, showing that Rosenjack and Deans joined the Knights in July 1909. They were given top jobs in the mine shortly after that, despite their young ages and a lack of experience and knowledge—Rosenjack was twenty-two years old, while Deans was twenty-one.

Rosenjack was a "cager," a hoisting engineer who raised and lowered men into the mines on a large iron cage. This was an engineering job—not a common job as a miner or laborer. Deans was an assistant cager. Assisting them was Matt Francisco, who was seventeen. These were higher paying jobs.

In order to have these jobs, one needed a state certificate of competency as a hoisting engineer. The annual coal reports from the state of Illinois list everyone with required certificates. Michele (Enrietta) Micetich at the Carbon Hill School Museum has an extensive set of these reports. The record shows that neither Rosenjack nor Deans were certified by the state.

Among those responsible for recruiting men for the Knights of Pythias in Cherry were Frank Buck, the mine company clerk, and Dr. Lyston D. Howe, the mine doctor. They were the ones who brought Rosenjack and Deans into the lodge, and who rewarded them with choice jobs for joining.

Mr. Rooney said there is no other reasonable explanation for these two young men getting such important jobs in the mine, and he believes it was Buck and Howe who made sure the young men got out of town quickly. If Rosenjack and Deans had testified before the coroner's jury, they likely would have been asked how young, inexperienced, and unqualified men got important jobs that carried such responsibility. They would have to admit it was a reward—a Chicago-style political pay-to-play way of doing business—for joining the lodge, boosting the membership for the benefit of Buck and Howe.

Some theories at the time suggested that Rosenjack and Deans left town because they may have feared for their safety or to avoid testifying to protect the mining company, and that the coal company paid their way. However, Mr. Rooney believes their avoidance was to protect Buck and Howe.

The record shows that Buck was hostile when testifying before the coroner's jury, being evasive, lying, and refusing to answer some questions.

Martin Powers, the checkweighman, testified at the inquest that Alex Norberg shouted at Rosenjack for his inept actions after the fire broke out. George Eddy did not equivocate; he told the coroner's jury that Rosenjack was to blame for the tragedy.

The coroner was frustrated in not hearing from the most important witnesses, saying Matt Francisco and Robert Deans "were the boys who actually set fire to the bale of hay," calling them "heedless boys" who "pushed the car of hay to an open torch that was dripping burning oil, and leave it there." They never had to answer the most important questions involved in this tragedy.

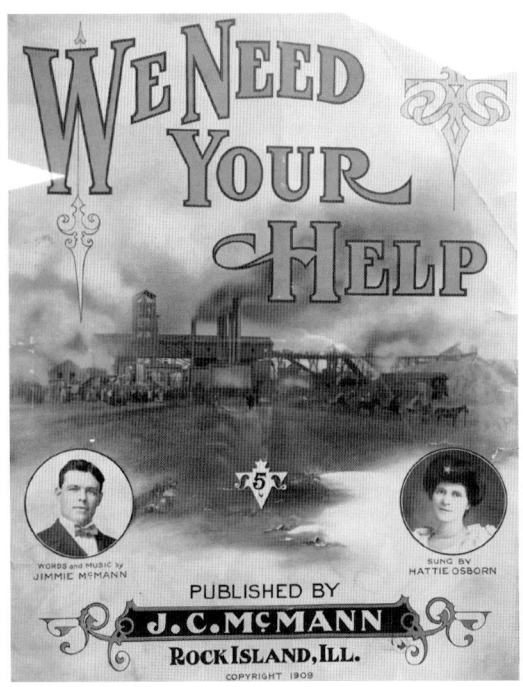

Appeals for help went out all over the world, in many forms, including the sheet music to this song.

9

THE MEMORIALS

Miners across Illinois contributed to a fund to build a monument to the Cherry workers who died. It was unveiled in Cherry Miner's Cemetery on November, 13, 1911, exactly two years after the disaster. A brass band led a solemn procession to the cemetery, just south of town. The Congregational Church was overflowing for its afternoon service. Politicians and union leaders gave speeches at the Miner's Hall.

Several thousand people attended the dedication of the 13-foot marble monument, which showed a grieving woman. The inscription reads: "In Memory of the Miners Who Lost Their Lives in the Cherry Mine Disaster, November 13, 1909. Erected by the U.M.W. of A., District No. 12, Illinois, Nov. 13, 1911."

John Thomas Brown was the last survivor of the twenty-one men who were rescued after being trapped in the mine for eight days in 1909. He attended the memorial in 1961, laying a wreath at the cemetery monument, along with the mayor of Cherry and several United Mine Workers officials. Brown had been living in Des Moines, Iowa, for a number of years, and did not know about the annual memorial services until then. It was believed that all the "Eight-Day Men" had died, until Brown was found. He participated in the memorial services for the first time in fifty-two years.

Brown (1885–1968) and his wife, Mary (1884–1949), were born in England. John came to America in 1909 to find work. He found it at the Cherry mine. He sent for Mary and their three children, Annie, Thomas and James, and they arrived on September 12, 1909, and headed for Cherry. James joined his father at the memorial services in 1961.

Several hundred people attended ceremonies on November 14 and 15, 2009, to commemorate the centennial of the Cherry mine tragedy. A new monument was dedicated in downtown Cherry, with the names of the dead engraved on two large marble stones. Among those telling their stories to a *LaSalle News-Tribune* reporter were descendants of Antenore Quartaroli and John Majersky. Representatives from the United Mine Workers and local politicians spoke, along with Alessandro Motta, counsel general of Italy, and Rita Cut, from the Emilia-Romagna area of Italy, the home area of most of the seventy-three Italian miners who died. Motta said the sorrow of the Cherry tragedy is still felt in Italy today. There were a number of special events, including trolley tours of the mine site and cemetery, displays in the school and the library, and services at the Cherry Miner's Memorial Cemetery.

The Illustrated History of the Cherry Mine Disaster of 1909

Dedication day, November 13, 1911.

The Memorials

Dedication of the memorial, November 13, 1911.

United Mine Workers representatives were among the large crowd at ceremonies observing the fiftieth anniversary in 1959.

The Illustrated History of the Cherry Mine Disaster of 1909

The United Mine Workers memorial stands between the adjoining Holy Trinity Miners Memorial Cemetery (Catholic) and Protestant Miners Memorial Cemetery, on the south edge of town.

The Memorials

A large exhibit, *The Flames Caught Us*, opened at the Abraham Lincoln Presidential Library and Museum in Springfield, Illinois, in 2009. The exhibit traveled to other locations after its Springfield run. It was more than 300 lineal feet of images, text, and diagrams on three levels.

At the opening of the exhibit, Michael Compasso of Dwight, Peggy Lami's cousin, was staring at the vinyl print of children waiting for milk. He turned to her and said, "That's your Auntie Ann and she's holding my dad."

It is an example of what happens when people get together to discuss a subject, and it is something that has happened often in the preparation of this book; people share information and everyone comes away with valuable knowledge they previously did not have.

The village of Cherry continues to commemorate the tragedy at the Cherry mine on its anniversary. Descendants of the mine victims from all over the country come to Cherry, not just for memorial observances, but all year long. Just south of town are Holy Trinity Miner's Cemetery and Protestant Miner's Cemetery, adjoining burial grounds which contain the bodies of many of the miners from the 1909 tragedy. The large monument dedicated in 1911 dominates the small cemetery.

The St. Paul Coal Company reopened the third shaft of the Cherry mine in late 1910. More than 250 men took jobs there. The mine operated until 1927. A great number of coal mines closed that year all across Illinois, after the industry changed from deep shaft mining to strip mining, where large machines stripped the coal veins at the surface. There were a number of strip mines in nearby Grundy, Will, and Kankakee counties, and more in central and southern Illinois. The last one in this area, in Essex, closed in 1974. A few men bought the Cherry mine in 1927 and mined coal on a small level until 1935.

The Cherry Library and Museum shares a building with City Hall. The new memorial plaza dedicated in 2009 is immediately to the south.

10

CHERRY TODAY

The small village of Cherry, Illinois, is still there. Its population, estimated at 2,500 at its peak, is approximately 500 today, in 2020. It is a village of beautiful houses and friendly people. The small downtown has a bank, a post office, a number of businesses, and the Cherry Public Library and Museum, which has an extensive collection of material, including numerous photographs, miners' tools, letters written by trapped miners, and more.

Today, all that remains of the Cherry mine are the "slag heaps" or "mine dumps," two large hills where the waste from the mining procedure was dumped. It stands in the Illinois cornfields. The vast area around the mine shafts have turned to farmland.

Governor George H. Ryan sent crews to the mine site in 2003 to erase a lot of history in what was called a "reclamation" project. The brick building that was part of the fan house was torn down. Steel girders were removed. The escape shaft, where the fire started, was buried. The wall of the boiler house was knocked down. Nothing was left but dirt. The mine hills were seeded with scrub brush. It made a lot of Cherry historians and preservationists angry. Historian Ray Tutaj, Jr., stated:

> It was like a knife through the heart … A total disrespect of history has been perpetrated here.
> No more steel girders standing in the prairie winds to remind those who visit that this was the location where miners entered the mine and where the twelve heroes came up on the fiery cage … There is no more escape shaft cover to see, for it is buried beneath a good heap of dirt. No more can we show people this was the exact spot where the fire began, 315 feet below.

Charles Bartoli, who owned the land, could do nothing to stop the state. Tutaj continued:

> They came in with their big trucks and starting resurfacing the landscape … The smaller original hills first created by the miners are leveled, too. The two large hills are still there.
> If you respect the history that has unfolded here, once upon a time, then you preserve what is left, you don't destroy it … Everywhere I look, another grain elevator is being torn down in the name of safety. I saw them tear down the Peterstown elevator which was on this Milwaukee Road Line between Ladd and Mendota. I saw first-hand how sturdy those things are built. The Cherry grain elevator was also torn down.

Cherry Today

A new memorial plaza was dedicated in 2009, on the centennial of the mine disaster. A carved stone monument of a coal car quotes a poem in the *Bureau County Record* from December 22, 1909. Behind it stands two triangle monuments with the names of each victim carved into it, and behind that is a carving of the mine tipple in stone. Bricks and benches are carved with names of donors to the project.

The Illustrated History of the Cherry Mine Disaster of 1909

Holy Trinity Catholic Church and United Church of Christ stand side by side today on Main Street.

11

THE VICTIMS

Here is the list of victims from the state. Some misspellings from the day are corrected as we now know them:

M. Adakosky, Foliani Agramanti, Joseph Alexius, Charles Amellani, Paul Amellani, Alfio Amidei, G. Atalakis, Peter Atsolopis, George Bakalar, Antone Barrozi, Mike Bastia, Milce Bauer, Frank Bawman, Lewis Bawman, Thomas Bayliff, J. Benossif, Charles Bernadini, Tonzothe Bertolioni, John Betot, Antonio Bolla, Peter Bolla, Joseph Bordesona, Adolph Bosviel, Jerome Boucher, Oliver Brain, Peter Bredenci, John Brown, Thomas Brown, Edward Bruno, John Bruzis, Richard Buckels, Charles Budzom, Joseph Budzon, John Bundy, Joseph Burke, Clemento Bursile, August Butilla, John Cagosky, Frank Camilli, Canivo Canov, Elfi Carlo, Dominick Casolari, Elizio Casollari, John Casserio, Chelsto Castoinelo, Chares Caviglini, John Celbulka, Joseph Chebubar, Peter Ciocci, Canical Cioci, Mike Cipola, Robert Clark, Henry Cohard, John Compasso, Henry Conlon, Angelo Costi, Lewis Costi, John Davies, Fred Demesey, Francisco Dinelli, Victor Detourney, John Donaldson, Andrew Dovin, George Dovin, Leopold Dumont, John Dunko, Benjamin Durand, Andrew Durdan, Miestre Elario (or Llirio), George Elko, Peter Eloses, Charles Erickson, Eric Ericsson, John Farlo, Peter Fayen, John Flood, John Forgach, Dominick Formento, August Franciskovic, John Franciskovic, Ole Freeburg, John Galletti, John Garabelda, John Garletti, Frank Geckse, Angone Giacolzza, Lewis Gibbs, John Governor, Andrew Grehaski, Frank Grumeth, Peter Gugliem, John Guidarini, Joseph Gulick, Jalindy Gwaltyeri, Steve Hadovski, August Hainant, Mike Halko, John Halofcak, Joseph Harpka, John Hertzel, Alfred Howard, Samuel Howard, John Hudar, William Hynds, Frank Jagodzinski, Frank James, James Jamison, Joe Janavizza, John Kanz, John Kenig, John Klaser, George Klemiar, Richard Klemiar, Thomas Klemiar, Dominick Kliklunas, John Kmetz, Antone Korvonia, Joseph Korvonia, Frank Kovocivio, Alex Kroll, Alfred Kroll, Henry Kroll, Julius Kussner, Paul Kutz, Frank Lallie, Frank Leadache, James Leadache, Joseph Leadache, John Liptak, Isaac Lewis, Urbain Leynaud, Charles Leyshon, John Lonzatti, Selcomo Lenzetti, David Love, James Love, John Love, Morrison Love, Andrew Lukachko, Mike Lurnas, John Maceoha, Joe Malinoski, Josip Malnar, Lewis Malnar, Archie Marchiona, Frank Marchiona, Anton Masenetta, William Matear, Frank Mayelemis, John Majersky, John

The Illustrated History of the Cherry Mine Disaster of 1909

Mazak, John Mazenetto, Robert McCandless, John McCrudden, Peter McCrudden, Andrew McFadden, John McGill, Andrew McLuckie, George McMullen, Joseph Meicora, Tonys Mekles, Arthur Merdior, Edward Miller, Arthur Mills, Edward Mills, John Mittle, Joseph Mokos, Joseph Monahan, Hasan Mumetich, Joseph Nanni, Alex Norberg, August Norberg, Charles Olson, Matt Onderko, Donaty Ossek, Martin Ossek, Andrew Packo, Albert Palmiori, Charles Papet, Anton Paulin, John Pavolski, Alex Pearson, John Pearson, Peter Perbacker, Dominick Perono, Ben Pete, Joseph Pressinger, Joseph Prich, Perys Prusitus, Peter Prusitus, John Pshak, Joe Raviso, Joseph Repsel, Martin Repsel, Cegu Ricca, Thomas Richards, Joseph Rimkus, Frank Rittel, Joseph Riva, Joseph Robeza, Joseph Rodonis, Victor Rolland, Robert Rossman, Gailamyo Ruggesie, Frank Ruygiesi, Olaf Sandeen, Julius Sarbelle, William Scotland, Edward Seitz, Paul Seitz, J. Semboa, John Sestak, Augusto Sergenti, Antone Shermel, John Shima, Andrew Siamon, John W. Smith, Cantina Sopko, James Speir, Antone Stam, Frank Stanchez, John Stark, Tony, Staszeski, James Stearns, Peter Steele, Dominick Stefenelli, Harry Stettler, Harry Stewart, Charles Sublich, John Suffen, John Suhe, Mike Suhe, Joseph Sukitus, John Szabrinski, Eugene Talioli, Pasquale Tamarri, Joseph Tamashanski, George Teszone, Andrew Timko, Joseph Timko, Sr., Joseph Timko, Jr., Steve Timko, Emilia Tonnelli, John Tonner, Frank Tosseth, Nocenti Turchi, Filippe Ugo, Charles Waite, Anthony Welkas, George White, William Wyatt, Frank Yacober, Peter Yannis, Joseph Yearly, Antone Yurcheak, Giatano Zacherria, Pat Zeikell, Joseph Zekula, and Thomas Zliegley.

Antonio Brassea died a few days after the disaster from burns and respiratory ailments caused by the fire. His name is not included in state's list of victims. He was fifty years old and came to America from Torino, Italy, in 1881.